T0296296

LONDON MATHEMATICAL SOCIETY LECTUR[E NOTE SERIES]

The books in the series listed below are available from booksellers, or, in case of difficulty, from Cambridge University Press.

4 Algebraic topology, J.F. ADAMS
17 Differential germs and catastrophes, Th. BROCKER & L. LANDER
27 Skew field constructions, P.M. COHN
34 Representation theory of Lie groups, M.F. ATIYAH *et al*
36 Homological group theory, C.T.C. WALL (ed)
39 Affine sets and affine groups, D.G. NORTHCOTT
40 Introduction to H_p spaces, P.J. KOOSIS
42 Topics in the theory of group presentations, D.L. JOHNSON
43 Graphs, codes and designs, P.J. CAMERON & J.H. VAN LINT
45 Recursion theory: its generalisations and applications, F.R. DRAKE & S.S. WAINER (eds)
46 p-adic analysis: a short course on recent work, N. KOBLITZ
49 Finite geometries and designs, P. CAMERON, J.W.P. HIRSCHFELD & D.R. HUGHES (eds)
50 Commutator calculus and groups of homotopy classes, H.J. BAUES
51 Synthetic differential geometry, A. KOCK
54 Markov processes and related problems of analysis, E.B. DYNKIN
57 Techniques of geometric topology, R.A. FENN
58 Singularities of smooth functions and maps, J.A. MARTINET
59 Applicable differential geometry, M. CRAMPIN & F.A.E. PIRANI
60 Integrable systems, S.P. NOVIKOV *et al*
62 Economics for mathematicians, J.W.S. CASSELS
65 Several complex variables and complex manifolds I, M.J. FIELD
66 Several complex variables and complex manifolds II, M.J. FIELD
68 Complex algebraic surfaces, A. BEAUVILLE
69 Representation theory, I.M. GELFAND *et al*
74 Symmetric designs: an algebraic approach, E.S. LANDER
76 Spectral theory of linear differential operators and comparison algebras, H.O. CORDES
77 Isolated singular points on complete intersections, E.J.N. LOOIJENGA
78 A primer on Riemann surfaces, A.F. BEARDON
79 Probability, statistics and analysis, J.F.C. KINGMAN & G.E.H. REUTER (eds)
80 Introduction to the representation theory of compact and locally compact groups, A. ROBERT
81 Skew fields, P.K. DRAXL
82 Surveys in combinatorics, E.K. LLOYD (ed)
83 Homogeneous structures on Riemannian manifolds, F. TRICERRI & L. VANHECKE
84 Finite group algebras and their modules, P. LANDROCK
85 Solitons, P.G. DRAZIN
86 Topological topics, I.M. JAMES (ed)
87 Surveys in set theory, A.R.D. MATHIAS (ed)
88 FPF ring theory, C. FAITH & S. PAGE
89 An F-space sampler, N.J. KALTON, N.T. PECK & J.W. ROBERTS
90 Polytopes and symmetry, S.A. ROBERTSON
91 Classgroups of group rings, M.J. TAYLOR
92 Representation of rings over skew fields, A.H. SCHOFIELD
93 Aspects of topology, I.M. JAMES & E.H. KRONHEIMER (eds)
94 Representations of general linear groups, G.D. JAMES
95 Low-dimensional topology 1982, R.A. FENN (ed)

96 Diophantine equations over function fields, R.C. MASON
97 Varieties of constructive mathematics, D.S. BRIDGES & F. RICHMAN
98 Localization in Noetherian rings, A.V. JATEGAONKAR
99 Methods of differential geometry in algebraic topology, M. KAROUBI & C. LERUSTE
100 Stopping time techniques for analysts and probabilists, L. EGGHE
101 Groups and geometry, ROGER C. LYNDON
103 Surveys in combinatorics 1985, I. ANDERSON (ed)
104 Elliptic structures on 3-manifolds, C.B. THOMAS
105 A local spectral theory for closed operators, I. ERDELYI & WANG SHENGWANG
106 Syzygies, E.G. EVANS & P. GRIFFITH
107 Compactification of Siegel moduli schemes, C-L. CHAI
108 Some topics in graph theory, H.P. YAP
109 Diophantine Analysis, J. LOXTON & A. VAN DER POORTEN (eds)
110 An introduction to surreal numbers, H. GONSHOR
111 Analytical and geometric aspects of hyperbolic space, D.B.A.EPSTEIN (ed)
112 Low-dimensional topology and Kleinian groups, D.B.A. EPSTEIN (ed)
113 Lectures on the asymptotic theory of ideals, D. REES
114 Lectures on Bochner-Riesz means, K.M. DAVIS & Y-C. CHANG
115 An introduction to independence for analysts, H.G. DALES & W.H. WOODIN
116 Representations of algebras, P.J. WEBB (ed)
117 Homotopy theory, E. REES & J.D.S. JONES (eds)
118 Skew linear groups, M. SHIRVANI & B. WEHRFRITZ
119 Triangulated categories in the representation theory of finite-dimensional algebras, D. HAPPEL
120 Lectures on Fermat varieties, T. SHIODA
121 Proceedings of *Groups - St Andrews 1985*, E. ROBERTSON & C. CAMPBELL (eds)
122 Non-classical continuum mechanics, R.J. KNOPS & A.A. LACEY (eds)
123 Surveys in combinatorics 1987, C. WHITEHEAD (ed)
124 Lie groupoids and Lie algebroids in differential geometry, K. MACKENZIE
125 Commutator theory for congruence modular varieties, R. FREESE & R. MCKENZIE
126 Van der Corput's method for exponential sums, S.W. GRAHAM & G. KOLESNIK
127 New directions in dynamical systems, T.J. BEDFORD & J.W. SWIFT (eds)
128 Descriptive set theory and the structure of sets of uniqueness, A.S. KECHRIS & A. LOUVEAU
129 The subgroup structure of the finite classical groups, P.B. KLEIDMAN & M.W.LIEBECK
130 Model theory and modules, M. PREST
131 Algebraic, extremal & metric combinatorics, M-M. DEZA, P. FRANKL & I.G. ROSENBERG (eds)
132 Whitehead groups of finite groups, ROBERT OLIVER

London Mathematical Society Lecture Note Series. 133

Linear Algebraic Monoids

Mohan S. Putcha
North Carolina State University

CAMBRIDGE UNIVERSITY PRESS
Cambridge
New York New Rochelle Melbourne Sydney

CAMBRIDGE UNIVERSITY PRESS
Cambridge, New York, Melbourne, Madrid, Cape Town, Singapore, São Paulo

Cambridge University Press
The Edinburgh Building, Cambridge CB2 8RU, UK

Published in the United States of America by Cambridge University Press, New York

www.cambridge.org
Information on this title: www.cambridge.org/9780521358095

First published 1988
Re-issued in this digitally printed version 2008

A catalogue record for this publication is available from the British Library

Library of Congress Cataloguing in Publication data

Putcha, Mohan S., 1952-
Linear algebraic monoids.
(London Mathematical Society lecture note series; 133)
Bibliography: p.
Includes index
1. Monoids.
I. Title. II. Series
QA169.P87 1988 512'.55 88-6103

ISBN 978-0-521-35809-5 paperback

CONTENTS

PREFACE

The purpose of this book is to present the subject matter of (connected) linear algebraic monoids. This subject has been developed in the last several years, primarily by Lex Renner and the author. The basic results have been obtained. The subject is now ripe for new developments and applications. It is with the hope of attracting new researchers to the subject that this book is being written.

The theory of linear algebraic monoids represents a rather beautiful blend of ideas from abstract semigroup theory, algebraic geometry and the theory of linear algebraic groups. For example, one of the first results of the author has been to show that the group of units is solvable if and only if the regular $\mathcal{J}$–classes of the monoid form a relatively complemented lattice (they always form a finite lattice). Equivalently the monoid is a semilattice of archimedean semigroups. These semigroups were abstractly characterized by the author in his undergraduate days. From the viewpoint of semigroup theory, (von–Neumann) regular semigroups represent the most important class of semigroups. Group theorists are generally most interested in reductive algebraic groups. Well, there is a connection. L. Renner and the author have shown that a connected algebraic monoid M with zero is regular if and only if the group of units is reductive. In this situation, the author has shown that the Tits building of the group of units can be described as the local semilattice of partial $\mathcal{J}$–class idempotent cross–sections of the monoid. Going in the converse direction, L. Renner and the author have shown that the biordered set (in the sense of Nambooripad) E of idempotents of M is completely determined by the Tits building of G and a type map λ from the finite lattice $\mathcal{U}$ of $\mathcal{J}$–classes of M into a finite Boolean lattice (the power set of the Dynkin diagram). Another indication of the beauty of the subject is Renner's generalization to algebraic monoids of the classical Bruhat decomposition for algebraic groups. Renner obtains his decomposition by simply replacing the Weyl group in the Bruhat decomposition by a certain finite fundamental inverse semigroup. For the general linear group, the Weyl group is of course the symmetric group. For the full matrix semigroup, Renner's semigroup is the symmetric inverse semigroup.

There are strong connections between algebraic monoids and certain compactifications of semisimple algebraic groups and homogeneous spaces being studied by DeConcini and Procesi [14], [15]. In this regard the classification theorem of Renner is crucial. Let G be a reductive group with a maximal torus T. Renner establishes a correspondence between connected normal algebraic monoids M with zero having G as the group of units and normal torus embeddings $T \hookrightarrow \overline{T}$ (with zero) on which the Weyl group action on T extends. Since normal torus embeddings have to do with rational polyhedral cones, this yields a discrete geometrical classification of normal connected regular monoids with zero. Renner establishes this classification by first proving a powerful extension theorem: For such monoids M, a homomorphism on G, extending to $\overline{T}$, extends to M.

For the most part we have included all proofs (in many cases simpler than the original), thereby making the book quite appropriate for reading by graduate students. There are a few exceptions. For example, the recent results of the author on conjugacy classes are stated and explained without proofs. However, enough examples are given to give the reader a good understanding. The same is done with a part of Renner's classification theorem.

NOTATION

Throughout this book, $\mathbb{Z}$, $\mathbb{Z}^+$, $\mathbb{R}$, $\mathbb{R}^+$, $\mathbb{Q}$, $\mathbb{Q}^+$ will denote the sets of all integers, all positive integers, all reals, all positive reals, all rationals, all positive rationals, respectively. If X,Y are sets then $X \setminus Y = \{x \in X \mid x \notin Y\}$. If $Y \subseteq X$, then $X \setminus Y$ will also be denoted by $\sim X$. We let $|X|$ denote the cardinality of X.

K will denote an algebraically closed field, which will remain fixed throughout this book. We let $K^* = K \setminus \{0\}$. If $x_1,\ldots,x_n$ are indeterminates, then $K[x_1,\ldots,x_n]$ will denote the commutative polynomial algebra in $x_1,\ldots,x_n$. If V is a vector space over K, then End(V) will denote the algebra of all linear transformations from V into V, GL(V) its group of units. We let $\mathcal{M}_n(K)$ denote the algebra of all $n \times n$ matrices over K, $K^n = K \times \ldots \times K$. If $A \in \mathcal{M}_n(K)$, then A^t, $\rho(A)$, det A will denote the <u>transpose</u> of A, <u>rank</u> of A and <u>determinant</u> of A, respectively. We further let

$$GL(n,K) = \{A \in \mathcal{M}_n(K) \mid \det A \neq 0\}$$
$$SL(n,K) = \{A \in \mathcal{M}_n(K) \mid \det A = 1\}$$
$$\mathcal{T}_n(K) = \{A \in \mathcal{M}_n(K) \mid A \text{ is upper triangular}\}$$
$$\mathcal{D}_n(K) = \{A \in \mathcal{M}_n(K) \mid A \text{ is diagonal}\}$$
$$\mathcal{T}_n^*(K) = \mathcal{T}_n(K) \cap GL(n,K)$$
$$\mathcal{D}_n^*(K) = \mathcal{D}_n(K) \cap GL(n,K)$$

If $A = (a_{ij}) \in \mathcal{M}_n(K)$, $B \in \mathcal{M}_p(K)$, then $A \otimes B = (a_{ij}B) \in \mathcal{M}_{np}(K)$, $A \oplus B =$

$\begin{bmatrix} A & 0 \\ 0 & B \end{bmatrix} \in \mathscr{M}_{n+p}(K)$.

Let $(P, \leq)$ be a partially ordered set. A subset Γ of P is a chain if for all $\alpha, \beta \in \Gamma$ either $\alpha \leq \beta$ or $\beta \leq \alpha$. If Γ is a finite chain, then the length of Γ is defined to be $|\Gamma| - 1$. If $\alpha, \beta \in P$, then α covers β if $\alpha > \beta$ and there is no $\gamma \in P$ with $\alpha > \gamma > \beta$. Let $\alpha, \beta \in P$. If α, β have a greatest lower bound, then this element is denoted by $\alpha \wedge \beta$ and is called the meet of α,β. If α,β have a least upper bound, then this element is denoted by $\alpha \vee \beta$ and is called the join of α,β. If $\alpha \wedge \beta$ exists for all $\alpha, \beta \in P$, then P is a $\wedge$–semilattice. If $\alpha \vee \beta$ exists for all $\alpha, \beta \in P$, then P is a $\vee$–semilattice. If P is both a $\wedge$–semilattice and a $\vee$–semilattice, then it is a lattice. A lattice P is complete if every subset has a least upper bound and a greatest lower bound in P. A lattice P with a maximum element 1 and a minimum element 0 is complemented if for all $\alpha \in P$ there exists $\alpha' \in P$ such that $\alpha \vee \alpha' = 1$, $\alpha \wedge \alpha' = 0$. A lattice P is relatively complemented if for all $\alpha, \beta \in P$ with $\alpha < \beta$, the interval $[\alpha,\beta] = \{\gamma \in P | \alpha \leq \gamma \leq \beta\}$ is complemented. A lattice, isomorphic to the lattice of all subsets of a set is called a Boolean lattice.

$\mathscr{B}(T), \mathscr{C}(T)$	Definitions 4.21, 9.9
$\beta(\Lambda), \beta^{-}(\Lambda), \xi(B), \xi^{-}(B)$	Definition 9.9
$\mathscr{H}, \mathscr{R}, \mathscr{L}, \mathscr{J}, \mathscr{D}, \mid$	Definition 1.1
$\mathscr{P}(B)$, $\mathscr{S}(B) = \mathscr{S}(\Lambda)$	Definition 10.14
$\mathscr{U}$	Definitions 1.5, 12.19, Chapter 14
$\mathscr{N}(E)$	Definition 12.6
$\tilde{\mathscr{U}}$	Definition 10.22
$\hat{\mathscr{U}}$	Definition 14.4
$\hat{E}, E_G, E_{\hat{\mathscr{U}}}$	Chapters 13, 14
$\mathscr{X}(G)$, $\mathscr{X}(M)$	Definition 4.17, Chapter 8
$\mathscr{Y}_r$, $\mathscr{Z}_r$	Definition 11.11
$\phi, \sigma_\alpha, G_\alpha, U_\alpha, T_\alpha (\alpha \in \phi)$	Definitions 4.43, 4.46

1 ABSTRACT SEMIGROUPS

As usual a set S with an associative operation is called a semigroup. If $\emptyset \neq X \subseteq S$, then $< X >$ will denote the subsemigroup of S generated by X and $E(X) = \{e \in X | e^2 = e\}$ the set of idempotents in X. If $e, f \in E(S)$, then $e \geq f$ if $ef = fe = f$. An equivalence relation σ on S is a congruence if for all $a,b,c \in S$, $a \sigma b$ implies $ac \sigma bc$, $ca \sigma cb$. If S' is a semigroup, then a map $\phi: S \to S'$ is a homomorphism if $\phi(ab) = \phi(a)\phi(b)$ for all $a, b \in S$. The corresponding congruence is called the kernel of ϕ. A bijection $*: S \to S$ is an involution if $(ab)^* = b^*a^*$, $(a^*)^* = a$ for all $a,b \in S$. A subsemigroup of S which is a group is called a subgroup. S is strongly π–regular ($s\pi r$) if for each $a \in S$, there exists $i \in \mathbb{Z}^+$ such that a^i lies in a subgroup of S. See [1], [19], [49]. If $a,b \in S$, then b is an inverse of a if $aba = a$, $bab = b$. An element $a \in S$ is regular if $axa = a$ for some $x \in S$, i.e. a has an inverse in S. S is regular if each element of S is regular. $\mathcal{M}_n(K)$ is a regular semigroup, and by the Fitting decomposition it is also an $s\pi r$–semigroup. A semigroup with an identity element is called a monoid. If S is a semigroup then $S^1 = S$ if S is a monoid, $S^1 = S \cup \{1\}$ with obvious multiplication if S is not a monoid. Let M be a monoid. An invertible element of M is called a unit. Let G denote the group of units of M. Then M is unit regular if for each $a \in M$, there exists $x \in G$ such that $a = axa$. Equivalently $M = E(M)G$. If M is unit regular, then any submonoid of M containing G is also unit regular.

Definition 1.1. Let S be a semigroup, a,b $\in$ S. Then

(i) a $\mathcal{R}$ b if ax = b, by = a for some x,y $\in$ S^1.

(ii) a $\mathcal{L}$ b if xa = b, yb = a for some x,y $\in$ S^1.

(iii) $\mathcal{D} = \mathcal{R} \circ \mathcal{L} = \mathcal{L} \circ \mathcal{R}$, $\mathcal{H} = \mathcal{R} \cap \mathcal{L}$.

(iv) a|b (a divides b) if xay = b for some x,y $\in$ S^1.

(v) a $\mathcal{J}$ b if a|b|a; $J_a = \{x \in S \mid a \mathcal{J} x\}$.

(vi) $J_a \geq J_b$ if a|b.

Remark 1.2. For S = $\mathcal{M}_n(K)$, $\mathcal{L}$, $\mathcal{R}$ are row equivalence and column equivalence, respectively. If a,b $\in$ S, then $J_a \geq J_b$ if and only if $\rho(a) \geq \rho(b)$.

Remark 1.3. Let S be a semigroup. Then

(i) $\mathcal{J}$, $\mathcal{R}$, $\mathcal{L}$, $\mathcal{H}$, $\mathcal{D}$ are equivalence relations called Green's relations. See [11], [24], [33] for details.

(ii) If a $\in$ S, then a lies in a subgroup of S if and only if a $\mathcal{H}$ e for some e $\in$ E(S). In such a case, the $\mathcal{H}$–class of a is the group of units of eSe.

(iii) If S′ is an sπr–subsemigroup of S, a $\in$ S′, e $\in$ E(S) and if a $\mathcal{H}$ e in S, then e $\in$ E(S′) and a $\mathcal{H}$ e in S′.

(iv) Let a,b,c $\in$ S. Then a $\mathcal{R}$ b implies ca $\mathcal{R}$ cb and a $\mathcal{L}$ b implies ac $\mathcal{L}$ bc.

(v) Let a $\in$ S, e $\in$ E(S), a $\mathcal{R}$ e, H the $\mathcal{H}$–class of e. Then Ha is the $\mathcal{H}$–class of a.

(vi) Let e,f $\in$ E(S). Then e $\mathcal{R}$ f if and only if ef = f, fe = e. Similarly e $\mathcal{L}$ f if and only if ef = e, fe = f.

(vii) Let a $\in$ S be regular. Then a = axa for some x $\in$ S. So e = ax, f = xa $\in$ E(S), e $\mathcal{R}$ a $\mathcal{L}$ f. Thus a $\in$ S is regular if and only if a $\mathcal{R}$ e for some e $\in$ E(S) if and only if a $\mathcal{L}$ f for some f $\in$ E(S).

(viii) Let D be a $\mathscr{D}$-class of S. Then an element of D is regular if and only if each element of D is regular. Let $a \in D$ be regular, x an inverse of a. Then $a \,\mathscr{R}\, ax \,\mathscr{L}\, x$. Hence $x \in D$.

The following well-known result is derived from Green [24], Miller and Clifford [48] and Munn [49].

Theorem 1.4. Let S be an sπr-semigroup, $a,b,c \in S$. Then

(i) $a \,\mathscr{J}\, ab$ implies $a \,\mathscr{R}\, ab$; $a \,\mathscr{J}\, ba$ implies $a \,\mathscr{L}\, ba$.

(ii) $ab \,\mathscr{J}\, b \,\mathscr{J}\, bc$ implies $b \,\mathscr{J}\, abc$.

(iii) If $e \in E(S)$, J, H the $\mathscr{J}$-class, $\mathscr{H}$-class of e, respectively, then $J \cap eSe = H$.

(iv) $\mathscr{J} = \mathscr{D}$ on S.

(v) $a \,\mathscr{J}\, a^2$ implies that the $\mathscr{H}$-class of a is a group.

(vi) $a \,\mathscr{J}\, ab \,\mathscr{J}\, b$ if and only if $a \,\mathscr{L}\, e \,\mathscr{R}\, b$ for some $e \in E(S)$; $a \,\mathscr{J}\, ba \,\mathscr{J}\, b$ if and only if $a \,\mathscr{R}\, e \,\mathscr{L}\, b$ for some $e \in E(S)$.

(vii) Any regular subsemigroup of S is an sπr-semigroup.

Proof. (i) Suppose $a \,\mathscr{J}\, ab$. Then $xaby = a$ for some $x,y \in S^1$. Then $x^i a(by)^i = a$ for all $i \in \mathbb{Z}^+$. There exists $j \in \mathbb{Z}^+$ such that $(by)^j \,\mathscr{H}\, e$ for some $e \in E(S)$. Then $a = ae \in a(by)^jS \subseteq abS$. Hence $a \,\mathscr{R}\, ab$.

(ii) By (i), $ab \,\mathscr{L}\, b$. So $abc \,\mathscr{L}\, bc \,\mathscr{J}\, b$.

(iii) If $a \in eSe \cap J$, then by (i), $e \,\mathscr{R}\, ea = a = ae \,\mathscr{L}\, e$. So $a \,\mathscr{H}\, e$.

(iv) Let $a, b \in S$ such that $a \,\mathscr{J}\, b$. Then there exist $x,y \in S^1$ such that $xay = b$. So $a \,\mathscr{J}\, xa \,\mathscr{J}\, xay = b$. By (i), $a \,\mathscr{L}\, xa \,\mathscr{R}\, b$. Hence $a \,\mathscr{D}\, b$.

(v) Let H denote the $\mathscr{H}$-class of a. By (i), $a^2 \,\mathscr{H}\, a$. So $a^2x = a$ for some $x \in S^1$. Then $a^{i+1}x^i = a$ for all $i \in \mathbb{Z}^+$. So $a^i \,\mathscr{R}\, a$ for all $i \in \mathbb{Z}^+$. By (i), $a^i \in H$ for all $i \in \mathbb{Z}^+$. There exists $j \in \mathbb{Z}^+$, $e \in E(S)$ such that $a^j \,\mathscr{H}\, e$. Then $e \in H$ and H is a group.

(vi) Suppose $a \mathscr{J} ab \mathscr{J} b$. Then by (i), $a \mathscr{R} ab \mathscr{L} b$. There exist $x, y \in S^1$ such that $abx = a$, $yab = b$. So $ya = yabx = bx$. Then $aya = a$, $bxb = b$. So $ya \in E(S)$, $a \mathscr{L} ya = bx \mathscr{R} b$. Conversely assume that there exists $e \in E(S)$ such that $a \mathscr{L} e \mathscr{R} b$. So $xa = by = e$ for some $x, y \in S$. Hence $ab | xaby = e | a | ab$. Thus $a \mathscr{J} ab$.

(vii) Let $a \in S'$. There exists $i \in \mathbb{Z}^+$, $e \in E(S)$ such that $b = a^i \mathscr{H} e$ in S. There exists $x \in S'$ such that $b^2xb^2 = b^2$. Then $bxb = e$. So $e \in E(S')$ and $b \mathscr{H} e$ in S'.

<u>Definition 1.5</u>. Let S be an $s\pi r$–semigroup. A $\mathscr{J}$–class J of S is <u>regular</u> if $E(J) \neq \emptyset$. Equivalently some (hence every) element of J is regular. Let $\mathscr{U} = \mathscr{U}(S)$ denote the partially ordered set of all regular $\mathscr{J}$–classes of S. If $J \in \mathscr{U}(S)$, then let $J^o = J \cup \{0\}$ with

$$a \circ b = \begin{cases} ab \text{ if } a, b, ab \in J \\ 0 \text{ otherwise} \end{cases}$$

Let S be a semigroup, $\emptyset \neq I \subseteq S$. Then I is a <u>right ideal</u> of S if $IS \subseteq I$; I is a <u>left ideal</u> of S if $SI \subseteq I$; I is an <u>ideal</u> of S if $S^1IS^1 \subseteq I$. The minimum ideal of S, if it exists, is called the <u>kernel</u> of S.

<u>Definition 1.6</u>. (i) A <u>completely simple semigroup</u> S is an $s\pi r$–semigroup with no ideals other than S.

(ii) A <u>completely 0–simple semigroup</u> S is an $s\pi r$–semigroup with 0, having no ideals other than $\{0\}$ and S, and having a non–zero idempotent.

<u>Remark 1.7</u>. (i) This is not the standard definition of completely simple or completely 0–simple semigroups. However this definition is equivalent to the standard one by Munn [49].

(ii) Let S be an $s\pi r$–semigroup, $J \in \mathscr{U}(S)$. If $a,b \in J$, then there exist $x,s,t \in S^1$ such that $sat = b$, $axa = a$. Then $b = (sax)a(xat) \in JaJ$. Thus J^0 is a completely 0–simple semigroup.

(iii) Let S be an $s\pi r$–semigroup, $J \in \mathscr{U}(S)$. If $E(J)^2 \subseteq J$, then by Theorem 1.4 (ii), $J^2 = J$ and hence J is completely simple.

(iv) A completely simple semigroup has only one $\mathscr{J}$–class while a completely 0–simple semigroup has two $\mathscr{J}$–classes.

Definition 1.8. Let G be a group, Γ, Λ non–empty sets.

(i) Let $P: \Lambda \times \Gamma \to G$ be any map. Let $S = \Gamma \times G \times \Lambda$ with $(i,g,j)(k,h,l) = (i,gP(j,k)h,l)$. Then S is a completely simple semigroup called a Rees matrix semigroup without zero over G (and sandwich map P).

(ii) Let $P: \Lambda \times \Gamma \to G \cup \{0\}$ be any map such that for all $i \in \Gamma$, there exists $j \in \Lambda$ such that $P(j,i) \neq 0$, for all $j \in \Lambda$ there exists $i \in \Gamma$ such that $P(j,i) \neq 0$. Let $S = (\Gamma \times G \times \Lambda) \cup \{0\}$ with

$$(i,g,j)(k,h,l) = \begin{cases} (i,gP(j,k)h,l) & \text{if } P(j,k) \neq 0 \\ 0 & \text{if } P(j,k) = 0 \end{cases}$$

Then S is a completely 0–simple semigroup, called a regular Rees matrix semigroup with zero over G (and sandwich map P).

The following result is due to D. Rees (see [11] or [33]).

Theorem 1.9. (i) Any completely simple semigroup is isomorphic to a Rees matrix semigroup without zero over a group.

(ii) Any completely 0–simple semigroup is isomorphic to a regular Rees matrix semigroup with zero over a group.

Proof. We prove (ii), since (i) follows from it. Let S be a completely 0–simple semigroup. Then $\mathcal{U}(S) = \{J,0\}$ where $J = S\backslash\{0\}$. Let $e \in E(J)$, H,R,L the $\mathcal{H}$–class, $\mathcal{R}$–class, $\mathcal{L}$–class of e, respectively. Let $\Gamma = L/\mathcal{R} = L/\mathcal{H}$, $\Lambda = R/\mathcal{L} = R/\mathcal{H}$. For $\lambda \in \Lambda$, choose $r_\lambda \in \lambda$, for $\gamma \in \Gamma$ choose $l_\gamma \in \gamma$. Let $\lambda \in \Lambda$, $\gamma \in \Gamma$. If $r_\lambda l_\gamma \in J$, then by Theorem 1.4 (i), $r_\lambda l_\gamma \in H$. Thus we have a map $P: \Lambda \times \Gamma \to H \cup \{0\}$ given by $P(\lambda,\gamma) = r_\lambda l_\gamma$. Let $\lambda \in \Lambda$. Then since r_λ is regular, there exists $f \in E(J)$ such that $r_\lambda \mathcal{L} f$. Since $\mathcal{J} = \mathcal{D}$, there exists $\gamma \in \Gamma$ such that $l_\gamma \mathcal{R} f$. By Theorem 1.4 (vi), $r_\lambda l_\gamma \neq 0$. Similarly for each $\gamma \in \Gamma$, there exists $\lambda \in \Lambda$ such that $r_\lambda l_\gamma \neq 0$. Let $S' = (\Gamma \times H \times \Lambda) \cup \{0\}$ be the Rees matrix semigroup with sandwich map P. Define $\psi: S' \to S$ as $\psi(0) = 0$, $\psi(\gamma,h,\lambda) = l_\gamma h r_\lambda$. Since $ehe = h$ for $h \in H$, we see by Theorem 1.4 that $l_\gamma \mathcal{R}\, l_\gamma h r_\lambda \mathcal{L}\, r_\lambda$. Let $h, h' \in H$, $\gamma \in \Gamma$, $\lambda \in \Lambda$ such that $l_\gamma h r_\lambda = l_\gamma h' r_\lambda$. There exist $y,z \in S$ such that $r_\lambda z = e = y l_\gamma$. It follows that $h = ehe = eh'e = h'$. Thus ψ is injective. That ψ is a homomorphism is immediate. So we need to show that ψ is surjective. Let $a \in J$. There exist $\gamma \in \Gamma$, $\lambda \in \Lambda$ such that $l_\gamma \mathcal{R}\, a\, \mathcal{L}\, r_\lambda$. There exist $y,z \in S$ such that $r_\lambda z = e = y l_\lambda$. Then $ya\, \mathcal{R}\, y l_\gamma = e$, $az\, \mathcal{L}\, r_\lambda z = e$. By Theorem 1.4, $ya\, \mathcal{R}\, yaz\, \mathcal{L}\, az$. Hence $h = yaz \in H$. Now $l_\gamma \mathcal{L}\, e\, \mathcal{R}\, y$ and so $f = l_\gamma y \in E(J)$, $f\, \mathcal{R}\, l_\gamma\, \mathcal{R}\, a$. So $l_\gamma ya = a$. Similarly $azr_\gamma = a$. Thus $l_\gamma h r_\lambda = a$. This proves the theorem.

Definition 1.10. Let S be a semigroup, $S = \bigcup_{\alpha\in\Omega} S_\alpha$ a partition of S into subsemigroups. Then S is a semilattice (union) of $S_\alpha (\alpha \in \Omega)$ if for all $\alpha,\gamma \in \Omega$ there exists $\delta \in \Omega$ such that $S_\alpha S_\gamma \cup S_\gamma S_\alpha \subseteq S_\delta$.

Definition 1.11. A semigroup S is completely regular if it is a union of its subgroups.

The following result is due to Clifford [10].

Theorem 1.12. A semigroup S is completely regular if and only if it is a semilattice of completely simple semigroups.

Definition 1.13. A semigroup S is archimedean if for all $a,b \in S$, $a|b^i$ for some $i \in \mathbb{Z}^+$.

The following result is due to Tamura and Kimura [114].

Theorem 1.14. Any commutative semigroup is a semilattice of archimedean semigroups.

The following result is due to the author [62]. The proof given here is due to Tamura [112].

Theorem 1.15. A semigroup S is a semilattice of archimedean semigroups if and only if for all $a,b \in S$, $a|b$ implies $a^2|b^i$ for some $i \in \mathbb{Z}^+$.

Proof. The necessity of the condition being obvious, assume that for all $a,b \in S$, $a|b$ implies $a^2|b^i$ for some $i \in \mathbb{Z}^+$. Then for all $a,b \in S$, $j \in \mathbb{Z}^+$, there exists $i \in \mathbb{Z}^+$ such that $a^j|b^i$. Define a relation η on S as follows: $a \,\eta\, b$ if $a|b^i$, $b|a^j$ for some $i,j \in \mathbb{Z}^+$. By the above, η is an equivalence relation on S. Let $a,b \in S$. Then $aba|(ab)^2$. So $a^2b|(aba)^2|(ab)^i$ for some $i \in \mathbb{Z}^+$. Continuing, we see that for all $j \in \mathbb{Z}^+$, there exists $k \in \mathbb{Z}^+$ such that $a^jb|(ab)^k$. Now let $a,b \in S$ such that $a \,\eta\, b$. Let $c \in S$. There exists $i \in \mathbb{Z}^+$ such that $a \mid b^i$. Then $xay = b^i$ for some $x,y \in S^1$. So $cxa|cb^i|(cb)^j$ for some $j \in \mathbb{Z}^+$. Hence $ac|(cxa)^2|(cb)^k$ for some $k \in \mathbb{Z}^+$. So $ac|(bc)^{k+1}$. Similarly $bc|(ac)^l$ for some $l \in \mathbb{Z}^+$. Thus $ac \,\eta\, bc$. Similarly $ca \,\eta\, cb$. Hence η is a congruence. Clearly $a \,\eta\, a^2$ for all $a \in S$. Let $a,b \in S$. Then $ab|(ba)^2$, $ba|(ab)^2$. Hence $ab \,\eta\, ba$. It follows that S is a semilattice of its η-classes. Let T be a η-class, $a,b \in T$. Then there exist $x,y \in S$ such that $xay = b^i$. Then $bxayb = b^{i+3}$ and $bx \,\eta\, xay \,\eta\, b \,\eta\, yxay \,\eta\, yb$. So $bx,yb \in T$ and $a|b^{i+3}$ in T. Thus T is an archimedean semigroup. This proves the theorem.

Let S be a semigroup, I an ideal of S. The Rees factor semigroup

$S/I = (S\backslash I) \cup \{0\}$ with

$$a \circ b = \begin{cases} ab & \text{if } ab \in S\backslash I \\ 0 & \text{otherwise} \end{cases}$$

If S/I is a nil semigroup, then S is a <u>nil extension</u> of I.

<u>Corollary 1.16</u>. Let S be an $s\pi r$–semigroup. Then the following conditions are equivalent.

(i) $E(J)^2 \subseteq J$ for all $J \in \mathscr{U}(S)$.

(ii) For all $a \in S, e \in E(S)$, $a \mid e$ implies $a^2 \mid e$.

(iii) S is a semilattice of archimedean semigroups.

(iv) S is a semilattice of nil extensions of completely simple semigroups.

<u>Proof</u>. (i) $\Rightarrow$ (ii). Let $a \in S, e \in E(S), a|e$. Then $xay = e$ for some $x,y \in S$. So $ayex, yexa \in E(J_e)$. Thus $(yexa)(ayex) \in J_e$ and $a^2|e$.

(ii) $\Rightarrow$ (iii). Let $a,b \in S$ such that $a|b$. Then $b^i \mathscr{H} e$ for some $e \in E(S), i \in \mathbb{Z}^+$. So $a \mid e$. Hence $a^2|e|b^i$.

(iii) $\Rightarrow$ (iv). Let S_α be an archimedean component of S. Let $a \in S_\alpha$. There exists $e \in E(S), n \in \mathbb{Z}^+$ such that $a^n \mathscr{H} e$ in S. So there exists $x \in S$ such that $a^n x = xa^n = e, ex = xe = x, ea^n = a^n e = a^n$. It follows that $e,x \in S_\alpha$. Hence S_α is an $s\pi r$–archimedean semigroup. It is obvious that an $s\pi r$–archimedean semigroup is a nil extension of a completely simple semigroup.

(iv) $\Rightarrow$ (i). Let $e,f \in E(S)$, $e \mathscr{J} f$. Then e,f lie in the same archimedean component. Therefore $e \mathscr{J} ef$.

<u>Corollary 1.17</u>. Let S be an $s\pi r$–semigroup which is a semilattice of archimedean semigroups, S' an $s\pi r$–subsemigroup of S. Then S' is a semilattice of

archimedean semigroups.

Proof. Let $J \in \mathcal{U}(S')$, $e,f \in E(J)$. Now $(efe)^i \mathcal{H} h$ for some $i \in \mathbb{Z}^+$, $h \in E(S')$. Then $e \geq h$, $e \mathcal{J} h$ in S. So $e = h$ by Theorem 1.4 (i). Hence $ef|e$ and $ef \in J$.

Definition 1.18. Let S, S' be semigroups, $\phi: S \to S'$ a homomorphism. Then ϕ is idempotent separating if ϕ is $1-1$ on $E(S)$. A congruence π on S is idempotent separating if for all $e,f \in E(S)$, $e \pi f$ implies $e = f$.

The following result is due to Lallement [40].

Proposition 1.19. Let S, S' be regular semigroups, $\phi: S \to S'$ a surjective homomorphism which is idempotent separating. Then

(i) $\phi(E(S)) = E(S')$

(ii) If $e,f \in E(S)$, then $\phi(e) \mathcal{R} \phi(f)$ implies $e \mathcal{R} f$; $\phi(e) \mathcal{L} \phi(f)$ implies $e \mathcal{L} f$; $\phi(e) \geq \phi(f)$ implies $e \geq f$.

(iii) If $a,b \in S$, then $\phi(a) = \phi(b)$ implies $a \mathcal{H} b$; $\phi(a) \mathcal{J} \phi(b)$ implies $a \mathcal{J} b$.

Proof. (i) Let $e' \in E(S')$. There exists $a \in S$ such that $\phi(a) = e'$. There exists $x \in S$ such that $a^2xa^2 = a^2$, $xa^2x = x$. Then $e = axa \in E(S)$, $\phi(e) = e'$.

(ii) Let $e,f \in E(S)$, $e' = \phi(e)$, $f' = \phi(f)$. Suppose $e'f' = f'$. Then there exists $x \in S$ such that $(ef)^2x(ef)^2 = (ef)^2$, $x(ef)^2x = x$. Let $f_1 = efx \in E(S)$. Then $\phi(f_1) = f' = \phi(f)$. So $f_1 = f$ and $ef = f$. Similarly $f'e' = f'$ implies $fe = f$.

(iii) Let $a,b \in S$, $\phi(a) = \phi(b)$. There exist $x,y \in S$ such that $axa = a$, $byb = b$. Let $e = ax$, $f = by \in E(S)$. Then $a \mathcal{R} e$, $f \mathcal{R} b$. So $\phi(e) \mathcal{R} \phi(a) = \phi(b) \mathcal{R} \phi(f)$. By (ii), $e \mathcal{R} f$. So $a \mathcal{R} b$. Similarly $a \mathcal{L} b$. Hence $a \mathcal{H} b$. The second statement is now immediate.

Definition 1.20. Let S be a regular semigroup. The congruence μ on S defined by: $a \mu b$ if and only if $xay \mathscr{H} xby$ for all $x,y \in S^1$ is called the fundamental congruence on S. S is said to be fundamental if μ is the equality on S.

Remark 1.21. Let S be a regular semigroup. Then

(i) By Proposition 1.19, μ is the largest idempotent separating congruence on S and S/μ is fundamental.

(ii) If $e \in E(S)$, then $\mu|eSe$ is the fundamental congruence on eSe. Let $a \in S$. Then $a \mu e$ if and only if $a \mathscr{H} e$ and $af = fa$ for all $f \in E(eSe)$. See Hall [31; Theorem 5].

(iii) If $S = \mathscr{M}_n(K)$, then μ is given by: $a \mu b$ if and only if $a = \alpha b$ for some $\alpha \in K^*$.

Definition 1.22. A semigroup S is an inverse semigroup if each $a \in S$ has a unique inverse, denoted by a^{-1}.

Remark 1.23. (i) A semigroup S is an inverse semigroup if and only if S is regular and $ef = fe$ for all $e,f \in E(S)$. In such a case $a \to a^{-1}$ is an involution of S. See [11], [33], [61].

(ii) Any commutative idempotent semigroup (called a semilattice) is an inverse semigroup.

(iii) If X is a set, then the semigroup $\mathscr{I}(X)$ of all partial one to one transformations on X is an inverse semigroup, called the symmetric inverse semigroup on X.

(iv) Let E be a semilattice and let T_E denote the subsemigroup of $\mathscr{I}(E)$ consisting of all isomorphisms $\alpha: eE \cong fE$ where $e,f \in E$. T_E is called the Munn semigroup of E.

(v) Let S be an inverse semigroup, $E = E(S)$. Let $a \in S$, $e = aa^{-1}$, $f = a^{-1}a$. Then $\alpha_a \in T_E$ where α_a: $eE \cong fE$ is given by $h\alpha_a = a^{-1}ha$. Then $\theta: S \to T_E$ given by $\theta(a) = \alpha_a$ is an idempotent separating homomorphism with kernel μ. Moreover μ is also given by: $a \mu b$ if and only if $a^{-1}ea = b^{-1}eb$ for all $e \in E(S)$. See [33; Section V] for details.

We therefore have the following result of Munn [50].

Theorem 1.24. Let E be a semilattice. Then T_E is a fundamental inverse semigroup with idempotent semilattice E. Moreover every fundamental inverse semigroup S with idempotent semilattice E is isomorphic to a subsemigroup of T_E containing E.

Remark 1.25. Let S be a regular semigroup, $E = E(S)$.

(i) Fitz-Gerald [23] (see also [21]) has shown that $\langle E \rangle$ is a regular semigroup. Hall [31] constructs a fundamental regular semigroup T_E and obtains an idempotent separating homomorphism $\theta: S \to T_{\langle E \rangle}$ with kernel μ.

(ii) The complete generalizations of the Munn representation to regular semigroups have been obtained by Grillet [27], [28] and Nambooripad [51], [52]. Grillet's approach has been to axiomatize the structures of the partially ordered sets $S/\mathcal{R}$ and $S/\mathcal{L}$ and the connections between them. Nambooripad's approach has been to introduce a biordered structure on E and to axiomatize it. It is this approach that will be most relevant to us. See Chapters 12, 13, 14.

(iii) For a class of regular semigroups called strongly regular Baer semigroups, Janowitz [36] obtains an equivalent of the fundamental representation. His paper precedes that of Munn [50]. See [57] for details.

2 ALGEBRAIC GEOMETRY

The algebraic geometry needed in this book is of a relatively elementary nature. We list in this chapter the needed results. In Chapter 16 we will need a few additional results. We refer to [32; Chapters IX, X], [108; Chapter 1] for details.

Definition 2.1. (i) $X \subseteq K^n$ is closed if it is the zero set of a collection of polynomials in $K[x_1,...,x_n]$. In such a case let $K[X] = K[x_1,...,x_n]/I$ where $I = \{f \in K[x_1,...,x_n] \mid f(X) = 0\}$.

(ii) A closed subset X of K^n is irreducible if it is not a union of two proper closed subsets. Equivalently $K[X]$ is an integral domain.

(iii) Let $X \subseteq K^n$, $Y \subseteq K^m$ be closed sets. Then a map $\phi = (\phi_1, ..., \phi_m): X \to Y$ is a morphism (or a polynomial map) if each $\phi_i \in K[X]$. In such a case, there is a natural K–homomorphism $\phi^*: K[Y] \to K[X]$. ϕ is dominant if $\overline{\phi(X)} = Y$. In such a case ϕ^* is injective.

Remark 2.2. (i) Hilbert's Nullstellensatz establishes a 1–1 correspondence between the closed subsets of K^n and the radical ideals of $K[x_1,...,x_n]$.

(ii) Hilbert's basis theorem states that every ideal of $K[x_1,...,x_n]$ is finitely generated. Thus the closed subsets of K^n satisfy the descending chain condition.

(iii) If $X \subseteq K^m$, $Y \subseteq K^n$ are closed sets, then $X \times Y$ is a closed subset of K^{m+n} and $K[X \times Y] \cong K[X] \otimes_K K[Y]$. Note that the topology on $X \times Y$ is not the product topology.

(iv) The topology on K^n is called the Zariski topology. It is not Hausdorff. However points are closed and every open cover has a finite subcover.

Definition 2.3. Let X be a topological space. Then

(i) X is irreducible if X is not a union of two proper closed subsets.

(ii) X is Noetherian if it satisfies the ascending chain condition on open sets (equivalently the descending chain condition on closed sets).

(iii) A subset Y of X is locally closed if it is open in its closure, i.e. Y is the intersection of an open subset and a closed subset of X.

(iv) A finite union of locally closed subsets of X is constructible.

Remark 2.4. (i) Any locally closed subspace of a Noetherian space is again Noetherian. In particular any locally closed subspace of K^n is Noetherian.

(ii) A Noetherian space X is uniquely expressible as a finite union of irreducible closed subsets, called the irreducible components of X.

(iii) If X is irreducible, U a non-empty open subset, then $\bar{U} = X$ and U is irreducible. In particular the intersection of two non-empty open subsets of X is again non-empty.

Definition 2.5. Let X be a topological space. Suppose that for each non-empty open set U of X, there is associated a K-algebra $\mathcal{O}(U)$ of K-valued functions of U such that with $\mathcal{O}(\phi) = \{0\}$, we have,

(i) If $\emptyset \neq U \subseteq V$ are open sets, $f \in \mathcal{O}(V)$, then $f|U \in \mathcal{O}(U)$.

(ii) Let U be a non-empty open set with an open covering $U_\alpha (\alpha \in \Gamma)$. Let $f: U \to K$ such that $f|U_\alpha \in \mathcal{O}(U_\alpha)$ for all $\alpha \in \Gamma$. Then $f \in \mathcal{O}(U)$.

Then $\mathcal{O} = \mathcal{O}_X$ is a sheaf of functions on X and $X = (X, \mathcal{O})$ is a ringed space.

Remark 2.6. (i) Let $X \subseteq K^n$ be a closed set. If U is a non-empty open subset of X, then let $\mathcal{O}(U) = \mathcal{O}_X(U) = \{\phi: U \to K \mid$ for all $x \in U$, there is an open subset V of U containing x and $f,g \in K[X]$ such that g is non-zero on V and $\phi = f/g$ on $V\}$. Then $X = (X,\mathcal{O})$ is a ringed space with $\mathcal{O}(X) = K[X]$.

(ii) Let $(X,\mathcal{O})$ be a ringed space, $Y \subseteq X$. Then $Y = (Y,\mathcal{O}')$ is the induced ringed space. Here Y is considered with the induced topology and $\mathcal{O}'$ is defined as follows. If $\emptyset \neq U$ is open in Y, then $\mathcal{O}'(U)$ consists of all functions $f: U \to K$ such that there is an open covering $U \subseteq \bigcup_{\alpha\in\Gamma} U_\alpha$ by open sets of X such that for each $\alpha \in \Gamma$, $f|U \cap U_\alpha = f_\alpha|U \cap U_\alpha$ for some $f_\alpha \in \mathcal{O}(U_\alpha)$. If U is open in X, then $\mathcal{O}'(U) = \mathcal{O}(U)$. See [108; Section 1.4] for details.

(iii) Let $Y \subseteq X$ be closed sets in K^n. Then the sheaf of functions on Y, given by (i), is that induced (as in (ii)) from the sheaf of functions on X.

Definition 2.7. An affine variety is a ringed space X isomorphic to a closed subset of some K^n. In such a case $\mathcal{O}(X)$ is denoted by $K[X]$ and called the affine algebra of X.

Remark 2.8. (i) A closed subset of an affine variety is again an affine variety.

(ii) Let X be an affine variety, $f \in K[X]$. Then $X_f = \{x \in X \mid f(x) \neq 0\} \cong \{(x,\alpha) \mid x \in X, \alpha \in K, f(x)\alpha = 1\}$ is affine.

(iii) $GL(n,K) = \{A \in \mathcal{M}_n(K) \mid \det A \neq 0\}$ is an affine variety.

(iv) $K^2\setminus\{0\}$ is not an affine variety.

(v) An open subset of an affine variety is a finite union of affine open subsets.

Definition 2.9. (i) Let X be an irreducible affine variety. Then the field of rational functions on X, $K(X)$ is defined to be the field of quotients of the integral domain $K[X]$. The dimension of X, $\dim X$ is defined to be the transcendental degree of

$K(X)$ over K.

(ii) If X is an affine variety, then the dimension of X, $\dim X$ is defined to be the maximum of the dimensions of the irreducible components of X.

Remark 2.10. Let X be an irreducible affine variety and U a non–empty affine open subset of X. Then $K(X) = K(U)$ and hence $\dim X = \dim U$.

Definition 2.11. (i) A prevariety is a Noetherian ringed space X which is covered by affine open subsets.

(ii) If X_1, X_2 are prevarieties, then $X = X_1 \times X_2$ is a prevariety with the topological and sheaf structure determined by the affine open sets $U_1 \times U_2$ where U_i is an affine open subset of X_i (see Remark 2.2 (iii)).

(iii) A prevariety X is a variety if the diagonal $\Delta = \{(x,x) \mid x \in X\}$ is closed in $X \times X$.

Remark 2.12. (i) Any locally closed subset of a variety is again a variety.

(ii) If a prevariety X has the property that any two points lie in an affine open subset of X, then it is a variety (see [34; Lemma 2.5]).

Definition 2.13. (i) Let X be an irreducible variety. Then the dimension of X, $\dim X$ is defined to be $\dim U$ where U is any non–empty affine open subset of X.

(ii) If X is any variety, then the dimension of X, $\dim X$ is defined to be the maximum of the dimensions of the irreducible components of X.

See [32], [34] or [108] for the following.

Proposition 2.14. Let X, Y be irreducible varieties. Then

(i) If $X' \neq X$ is a non–empty closed subset of X, then $\dim X' < \dim X$.

(ii) If U is a non-empty open subset of X, then $\bar{U} = X$ and $\dim U = \dim X$.

(iii) $X \times Y$ is an irreducible variety and $\dim X \times Y = \dim X + \dim Y$.

The affine variety K^n is called the <u>affine</u> n-<u>space</u> and is usually denoted by A^n. The <u>projective</u> n-<u>space</u> P^n is defined as follows: $P^n = (K^{n+1}\backslash\{0\})/\sim$ where for $a,b \in K^{n+1}\backslash\{0\}$, $a \sim b$ if $b = \alpha a$ for some $\alpha \in K^*$. A subset X of P^n is <u>closed</u> if it is the zero set of a collection of homogeneous polynomials in $K[x_o,...,x_n]$. If U is a non-empty open subset of P^n, then a function $\phi: U \to K$ is <u>regular</u> if for all $a \in U$, there exists an open subset V of U containing a and homogeneous polynomials $f,g \in K[x_o,...,x_n]$ of the same degree such that g is non-zero on V and $\phi = f/g$ on V. Let $\mathcal{O}(U)$ denote the K-algebra of all K-valued regular functions on U. Thus $P^n = (P^n, \mathcal{O})$ is a ringed space. Let $\alpha = (\alpha_o,...,\alpha_n) \in K^{n+1}\backslash\{0\}$. Let $U'_\alpha = \{(a_o,...,a_n) \in K^{n+1}\backslash\{0\} \mid \Sigma\alpha_i a_i \neq 0\}$, $U_\alpha = U'_\alpha/\sim$. Clearly $U_\alpha \cong \{(b_o,...,b_n) \mid b_o,...,b_n \in K, \Sigma\alpha_i b_i = 1\}$ is affine. Any two points in P^n lie in some U_α. It follows that P^n is an irreducible variety of dimension n.

<u>Definition 2.15</u>. (i) A variety isomorphic to a closed subset of some P^n is called a <u>projective</u> variety.

(ii) A variety isomorphic to a locally closed subset of some P^n is called a <u>quasi-projective variety</u>.

<u>Remark 2.16</u>. (i) Any affine variety is quasi-projective.

(ii) Let $S = \mathcal{M}_2(K)$, $e = \begin{bmatrix} 1 & 0 \\ 0 & 0 \end{bmatrix}$. Let R, J denote the $\mathcal{R}$-class, $\mathcal{J}$-class of e, respectively. Then $J/\mathcal{L} \cong R/\mathcal{H} \cong P^1$. Also $(S/\mu)\backslash\{0\} \cong P^3$.

<u>Definition 2.17</u>. Let X,Y be varieties. A map $\phi: X \to Y$ is a <u>morphism</u> if for any open set V of Y and $f \in \mathcal{O}(V)$, $U = \phi^{-1}(V)$ is open in X and $f \circ \phi \in \mathcal{O}(U)$. ϕ is <u>dominant</u> if $\overline{\phi(X)} = Y$.

Remark 2.18. (i) The restriction of a morphism to a subvariety is again a morphism.

(ii) For affine varieties X, Y, this definition agrees with that given in Definition 2.1 (iii).

(iii) If X, Y are varieties, $U_\alpha (\alpha \in \Gamma)$ an open covering of X, then a map $\phi: X \to Y$ is a morphism if and only if $\phi | U_\alpha$ is a morphism for all $\alpha \in \Gamma$.

We refer to [34; Theorem 4.4] for the following

Theorem 2.19. Let $\phi: X \to Y$ be a morphism of varieties. Then for any constructible subset V of X, $\phi(V)$ is constructible in Y. In particular, if X is irreducible, $\phi(X)$ contains a non-empty open subset of $\overline{\phi(X)}$.

See [34; Theorem 6.2] for the following.

Theorem 2.20. Let Y be a projective variety, X any variety. Then the map $\phi: X \times Y \to X$ given by $\phi(x,y) = x$ is a closed morphism.

See [17; Section 4.5, Theorem 2] or [32; Theorems 2.1, 4.3] for the following dimension theorem.

Theorem 2.21. Let $\phi: X \to Y$ be a dominant morphism of irreducible varieties, $\dim X = n$, $\dim Y = m$. Then

(i) $m \leq n$.

(ii) For any closed irreducible subset W of Y and any irreducible component V of X with $\phi(V) = W$, we have $\dim V \geq \dim W + n-m$.

(iii) There exists a non-empty open subset U of Y contained in $\phi(X)$ such that for any closed irreducible subset W of Y with $W \cap U \neq \emptyset$ and any irreducible component V of $\phi^{-1}(W)$ with $\phi(V) = W$, we have $\dim V = \dim W + n-m$.

3 LINEAR ALGEBRAIC SEMIGROUPS

The subject matter of this book is linear algebraic semigroups. We are now in a position to define this concept.

Definition 3.1. A (linear) algebraic semigroup $S = (S,o)$ is an affine variety S along with an associative product map, $o: S \times S \to S$ which is also a morphism of varieties. A homomorphism between algebraic semigroups S, S' is a semigroup homomorphism $\phi: S \to S'$ which is also a morphism of varieties. Isomorphisms and involutions are similarly defined.

Remark 3.2. If S is an algebraic semigroup, $e \in E(S)$, then $eS, Se, \overline{SeS}$ are closed subsemigroups of S and hence algebraic semigroups.

Example 3.3. Any finite dimensional algebra over K with respect to multiplication or the circle operation: $a \circ b = a+b-ab$ is an algebraic semigroup.

Example 3.4. If S is any (multiplicative) subsemigroup of $\mathcal{M}_n(K)$, then $\overline{S}$ is an algebraic semigroup.

Example 3.5. Any finite semigroup is an algebraic semigroup.

Example 3.6. Let X be a closed subset of K^n. Then $S = \{A \mid A \in \mathcal{M}_n(K), XA \subseteq X\}$ is a closed subsemigroup of $\mathcal{M}_n(K)$.

Example 3.7. Let $P \in \mathcal{M}_n(K)$ and let $S_1 = \{A \mid A \in \mathcal{M}_n(K), A^tPA = 0\}$, $S_2 = \{A \mid A \in \mathcal{M}_n(K), A^tPA = P\}$. Then S_1 is a closed subsemigroup of $\mathcal{M}_n(K)$, $0 \in S_1$ and S_2 is a closed submonoid of $\mathcal{M}_n(K)$.

Example 3.8. Two trivial algebraic semigroups on any affine variety X: (i) $xy = y$ for all $x, y \in X$, (ii) $xy = u$ for all $x,y \in X$, u a fixed element of X.

Example 3.9. (Rees construction). Let S be an algebraic semigroup, X,Y affine varieties, $\phi: Y \times X \to S$ a morphism. Let $\hat{S} = X \times S \times Y$ with

$$(x,s,y)(x', s', y') = (x,s\phi(y,x')s', y')$$

Then $\hat{S}$ is an algebraic semigroup.

Example 3.10. The map $\phi: \mathcal{M}_n(K) \oplus \mathcal{M}_p(K) \to \mathcal{M}_{np}(K)$ given by $\phi(A \oplus B) = A \otimes B$ is a homomorphism.

Example 3.11. The map $\phi: \mathcal{M}_n(K) \to \mathcal{M}_{n^2}(K)$ given by $\phi(A) = A \otimes A$ is an idempotent separating homomorphism.

Example 3.12. Let $S = \mathcal{M}_n(K)$. Consider the homomorphism $\phi: S \to S$ given by $\phi(a) = (\det a)a$. Then $\phi(S)$ is not closed in S, $\overline{\phi(S)} = S$.

Example 3.13. (Semidirect Product). Let S_1,S_2 be algebraic semigroups. Suppose for $a \in S_1$, $b \in S_2$, an element $a^b \in S_1$ is uniquely determined. Suppose that the map $(a,b) \to a^b$ is a morphism and that for all $a,a_1,a_2 \in S_1$, $b,b_1,b_2 \in S_2$,

$$(a_1a_2)^b = a_1^ba_2^b, \quad a^{b_1b_2} = (a^{b_2})^{b_1}$$

Let $S = S_1 \times S_2$ with multiplication

$$(a_1,b_1)(a_2,b_2) = (a_1a_2^{b_1},b_1b_2)$$

Then S is an algebraic semigroup, called the semidirect product of S_1,S_2.

Problem 3.14. Generalize the Krohn–Rhodes decomposition theorem for finite semigroups [39] to linear algebraic semigroups. The more recent approach of Rhodes [103] might also be relevant

The following result is well–known [16].

Theorem 3.15. Let M be a linear algebraic monoid. Then M is isomorphic to a closed submonoid of some $\mathcal{M}_n(K)$.

Proof. We may assume that M is a closed subset of some K^d. Since the operation on M is polynomially defined, there exist morphisms $f_1,\dots,f_m$ from M into K^d, $g_1,\dots,g_m \in K[S]$ such that for all $a,b \in M$,

$$ba = \sum_{i=1}^{m} g_i(a)f_i(b)$$

Let V denote the vector space of all maps from S into K^d. If $h \in V$, $a,x, \in M$, let $h_a(x) = h(xa)$. If $a \in M$, $h \in V$, let $\Gamma_a(h) = h_a$. Then $\Gamma_a \in \text{End}(V)$ and $\Gamma_a \circ \Gamma_b = \Gamma_{ab}$ for all $a,b \in M$. Let $I \in V$ denote the identity map. Then for all $x \in M$,

$I_a(x) = xa = \sum_{i=1}^{m} g_i(a)f_i(x)$. So

$$I_a = \sum_{i=1}^{m} g_i(a)f_i$$

Thus I_a is in the span of $f_1,\dots,f_m$. Let Y denote the finite dimensional space spanned by $I_a (a \in M)$. Then each element of Y is a morphism from M into K^d. If $a,b \in M$, then $\Gamma_a(I_b) = I_{ab}$. Hence $\Gamma_a(Y) \subseteq Y$. If $a \in M$, let $\Gamma(a)$ denote the restriction of Γ_a to Y. Then $\Gamma\colon M \to \mathrm{End}(Y)$ is a monoid homomorphism. There exist $1 = a_1,\dots,a_n \in M$ such that $w_i = I_{a_i}$, $i = 1,\dots,n$ form a basis of Y. Then

$$\Gamma_a(w_j) = I_{aa_j} = \sum_{i=1}^{m} g_i(aa_j)f_i = \sum_{i=1}^{m} h_{ij}(a)f_i \tag{1}$$

where $h_{ij}(a) = g_i(aa_j)$. Clearly each h_{ij} is a morphism on M. Extend $w_1,\dots,w_n$ to a basis $w_1,\dots,w_q$ of $\langle f_1,\dots,f_m\rangle$. Let

$$f_i = \sum_{k=1}^{q} \alpha_{ki}w_k \quad , \quad i = 1,\dots,m$$

Then by (1),

$$\Gamma_a(w_j) = \sum_{k=1}^{q}\sum_{i=1}^{m} \alpha_{ki}h_{ij}(a)w_k \quad , j = 1,\dots,n$$

Let $u_{jk}(a) = \sum_{i=1}^{m} \alpha_{ki}h_{ij}(a)$. So u_{jk} is a morphism on M and $\Gamma_a(w_j) = \sum_{k=1}^{q} u_{jk}(a)w_k$,

$j = 1,\ldots,n$. But $\Gamma_a(w_j) \in Y$. So

$$\Gamma_a(w_j) = \sum_{k=1}^{n} u_{jk}(a)w_k \quad , \quad j = 1,\ldots,n$$

Then $\phi(a) = (u_{jk}(a))^t \in \mathcal{M}_n(K)$ is the matrix of Γ_a. So $\phi\colon M \to \mathcal{M}_n(K)$ is a homomorphism of algebraic monoids. Now for all $a \in M$,

$$\sum_{k=1}^{n} u_{1k}(a)a_k = \sum_{k=1}^{n} u_{1k}(a)w_k(1) = \Gamma_a(w_1)(1) = a \tag{2}$$

If $A = (\beta_{kj}) \in \mathcal{M}_n(K)$, let $\psi(A) = \sum_{k=1}^{n} \beta_{k1}a_k \in K^d$. Then ψ is a morphism and by (2), $\psi(\phi(a)) = a$ for all $a \in M$. So $\phi(M) = \{A \in \mathcal{M}_n(K) \mid \psi(A) \in M, \phi(\psi(A)) = A\}$ is closed in $\mathcal{M}_n(K)$ and $\psi = \phi^{-1}$ on $\phi(M)$. This proves the theorem.

Corollary 3.16. Let S be an algebraic semigroup. Then S is isomorphic to a closed subsemigroup of some $\mathcal{M}_n(K)$.

Proof. We may assume that $S = (S,\circ)$ is a closed subset of some K^d. Let $u \in S$ and set $M = (S \times \{0\}) \cup \{(u,1)\} \subseteq K^{d+1}$, $S' = S \times \{0\} \subseteq M$. On M define

$$(a,\alpha)\circ(b,\beta) = ((1-\alpha)(1-\beta)(a\circ b) + (\alpha+\beta-\alpha\beta)(a+b-u),\alpha\beta)$$

Then $M = (M,\circ)$ is an algebraic monoid, $S \cong S' \subseteq M$. We are now done by Theorem 3.15.

Remark 3.17. Let M be a closed submonoid of $\mathcal{M}_n(K)$. Suppose M has a zero e, $\rho(e) = r$. If $a \in M$, let $\phi(a) = a-e$. Then $\phi: M \cong \phi(M) \subseteq (1-e)\mathcal{M}_n(K)(1-e) \cong \mathcal{M}_{n-r}(K)$. Note that e corresponds to the zero of $\mathcal{M}_{n-r}(K)$.

The following result was pointed out to the author by W. E. Clark (see [64; Corollary 1.4]).

Theorem 3.18. Let S be a closed subsemigroup of $\mathcal{M}_n(K)$. Then for all $a \in S$, a^n lies in a subgroup of S.

Proof. Let $a \in S$, $b = a^n$. By the Fitting decomposition, $b \mathcal{H} e$ for some $e = e^2 \in \mathcal{M}_n(K)$. Now $S_1 = \{x \in S \mid ex = xe = x\}$ is a closed subsemigroup of S and $b \in S_1$. There exists $c \in \mathcal{M}_n(K)$ such that $ec = ce = c$, $bc = cb = e$. Now for all $i \in \mathbb{Z}^+$, $b^iS_1 = \{x \in S_1 \mid c^ix \in S_1\}$ is closed and

$$bS_1 \supseteq b^2S_1 \supseteq \ldots.$$

Hence $b^iS_1 = b^{i+1}S_1$ for some $i \in \mathbb{Z}^+$. Then $S_1 = eS_1 = c^ib^iS_1 = c^ib^{i+1}S_1 = ebS_1 = bS_1$. Similarly $S_1 = S_1b$. There exists $x \in S_1$ such that $b = bx$. So $x = ex = cbx = cb = e$. Hence $e \in E(S_1)$. There exist $y,z \in S_1$ such that $by = e = zb$. It follows that $b \mathcal{H} e$ in S.

Corollary 3.19. A closed subsemigroup of $\mathcal{D}_n(K)$ is a semilattice of groups.

Corollary 3.20. A closed subsemigroup S of $\mathcal{T}_n(K)$ is a semilattice of archimedean semigroups. In particular $J^2 = J$ is completely simple for all $J \in \mathcal{U}(S)$.

Proof. By Corollaries 1.16, 1.17, Theorem 3.18, we may assume that $S = \mathcal{T}_n(K)$. If $a \in S$, let $\phi(a)$ denote the diagonal matrix with the same diagonal as a. Then

$\phi: S \to \mathscr{D}_n(K)$ is a homomorphism. Let $J \in \mathscr{U}(S)$, $e_1, e_2 \in E(J)$. Then $\phi(e_1) \mathscr{J} \phi(e_2)$ in $\mathscr{D}_n(K)$. Since $\mathscr{D}_n(K)$ is commutative, $\phi(e_1) = \phi(e_2)$. So $e_2e_1 - e_1$ is nilpotent. So $u = 1 + e_2e_1 - e_1 \in \mathscr{T}_n^*(K)$. Clearly $e_1u = e_1e_2e_1$. Thus $\rho(e_1) = \rho(e_1e_2e_1)$ and $e_1 \mathscr{H} e_1e_2e_1$ in $\mathscr{M}_n(K)$ and hence in S by Remark 1.3 (iii). Thus $e_1e_2 \in J$ and $E(J)^2 \subseteq J$. We are done by Corollary 1.6.

Remark 3.21. Corollary 3.20 is clearly valid for any $s\pi r$–subsemigroup of $\mathscr{T}_n(K)$. Thus any regular subsemigroup of $\mathscr{T}_n(K)$ is completely regular.

Definition 3.22. If e is an idempotent in $\mathscr{M}_n(K)$, $a \in \mathscr{M}_n(K)$, then $\det_e(a) = \det(eae + 1-e)$. If Γ is a finite set of idempotents in $\mathscr{M}_n(K)$, then $\det_\Gamma(a) = \prod_{f\in\Gamma} \det_f(a)$.

Remark 3.23. Let S be a closed subsemigroup of $\mathscr{M}_n(K)$, $e \in E(S)$, $a \in S$. Then by Theorems 1.4, 3.18, $\det_e(a) \neq 0$ if and only if $eae \mathscr{H} e$ in S if and only if $eae \mathscr{J} e$ in S.

Definition 3.24. An algebraic semigroup which is also a group is an algebraic group.

Remark 3.25. Let G be an algebraic group. Then by Theorem 3.15, G is a closed subgroup of some $GL(n,K)$. So $a^{-1} = (1/\det a) \operatorname{adj} a$ for all $a \in G$, where adj a denotes the adjoint of a. Hence the map $a \to a^{-1}$ is a morphism on G.

Corollary 3.26. Let S be an algebraic semigroup, $e \in E(S)$, H the $\mathscr{H}$–class of e. Then H is an algebraic group.

Proof. $H = \{a \in eSe \mid \det_e(a) \neq 0\} \cong G = \{(a,\alpha) \mid a \in eSe, \alpha \in K, \alpha \det_e(a) = 1\}$ is an algebraic group.

Lemma 3.27. Let S be an algebraic semigroup, $e \in E(S)$. Then the ideal $I = \{a \mid a \in S, a \nmid e\}$ is closed in S.

Proof. Let H denote the $\mathcal{H}$-class of e. By Remark 3.23, $X = eSe\backslash H$ is closed in S. Let $a \in I$. Then $exaye \in X$ for all $x,y \in X$. Conversely let $a \in S$ such that $exaye \in X$ for all $x,y \in S$. We claim that $a \in I$. Suppose not. Then $xay = e$ for some $x,y \in S$. So $exaye = e \in H$, a contradiction. Thus $I = \{a \in S \mid exaye \in X$ for all $x,y \in S\}$ is closed.

The following result is due to the author [64].

Theorem 3.28. Let S be an algebraic semigroup. Then $\mathcal{U}(S)$ is a finite partially ordered set. In particular S has a kernel.

Proof. Suppose the theorem is false. Then there exists an infinite set $X \subseteq E(S)$ such that for all $e,f \in X$, $e \mathcal{J} f$ implies $e = f$. For $e \in X$, let $I(e) = \{a \mid a \in S, a \nmid e\}$ which is closed by Lemma 3.25. We claim that there exists an infinite subset Y of X such that $|I(e) \cap Y| < \infty$ for all $e \in Y$. Suppose not. Then there exists $f_1 \in X$ such that $X_1 = I(f_1) \cap X$ is infinite. There exists $f_2 \in X$ such that $X_2 = I(f_2) \cap X_1$ is infinite. Continuing, we find a sequence $f_1, f_2, \ldots$ in X such that $f_{i+1} \in X_i = I(f_1) \cap \ldots \cap I(f_i) \cap X$ for all $i \in \mathbb{Z}^+$. Then $f_{i+1} \in I(f_1) \cap \ldots \cap I(f_i)$, $f_{i+1} \notin I(f_{i+1})$ for all $i \in \mathbb{Z}^+$. So we have a strictly descending chain of closed sets,

$$I(f_1) \supsetneq I(f_1) \cap I(f_2) \supsetneq I(f_1) \cap I(f_2) \cap I(f_3) \supsetneq \ldots$$

This contradiction shows that there exists an infinite subset Y of X such that $I(e) \cap Y$ is finite for all $e \in Y$. Choose $e_1 \in Y$. There exists $e_2 \in Y\backslash I(e_1)$ such that $e_1 \neq e_2$. Similarly there exists $e_3 \in Y\backslash(I(e_1) \cup I(e_2))$ such that $e_1 \neq e_3$, $e_2 \neq e_3$. Thus we find distinct idempotents $e_1, e_2, \ldots$ in Y such that $e_i \mid e_j$, $e_i \not\mathcal{J} e_j$ for $i < j$.

Since $e_1|e_2$, there exist $x,y \in S$ such that $xe_1y = e_2$. Let $e'_2 = e_1ye_2xe_1 \in E(S)$. Then $e_2 \not\mathcal{J} e'_2$, $e_1 \geq e'_2$. So $e_1 \neq e'_2$. Continuing, we find a sequence of idempotents, $e_1 > e'_2 > e'_3 > \ldots$ in S. This is a contradiction since S is a matrix semigroup.

Remark 3.29. When S is an idempotent semigroup, the above result has also been obtained by Sizer [107]. Theorem 3.28 implies that S has ideals $I_o \subseteq \ldots \subseteq I_m = S$, such that I_o is the completely simple kernel of S and each Rees factor semigroup I_k/I_{k-1}, $k = 1,\ldots,m$ is either nil or completely 0–simple. See [64]. Kleiman [38] has shown that the ideals I_k can be chosen to be closed. For generalizations of Theorem 3.28 to $s\pi r$–matrix semigroups see the author [64], [81] and Okninski [59]. That an $s\pi r$–matrix semigroup has a kernel is an early result of Clark [8].

The following result is due to the author [65].

Corollary 3.30. Let S be a closed subsemigroup of $\mathcal{M}_n(K)$, I an ideal of S. Then

(i) $\hat{I} = \{a \in S \mid a^n \in I\}$ is closed in S.

(ii) $\overline{I}/I$ is a nil semigroup and $a^n \in I$ for all $a \in \overline{I}$.

Proof. We prove only (i), since (ii) follows from it. If $e \in E(S)$, let $X(e) = \{a \in S \mid a^n \in SeS\}$. Let $a \in \hat{I}$. Then $a^n \in I$. By Theorem 3.16, $a^n \mathcal{H} e$ for some $e \in E(S)$. Then $e \in a^nS \subseteq I$. So $\hat{I} = \bigcup_{e \in E(I)} X(e)$. By Theorem 3.26, the family $\{SeS \mid e \in E(S)\}$ and hence the family $\{X(e) \mid e \in E(S)\}$ is finite. Thus we are reduced to showing that $X(e)$ is closed for all $e \in E(S)$. Fix $e \in E(S)$. For $f \in E(S)$, let $I(f) = \{a \in S \mid a \nmid f\}$. Then $I(f)$ is closed by Lemma 3.25. Let $F = \{f \mid f \in E(S), e \nmid f\}$. Then $SeS \subseteq I(f)$ for all $f \in F$. So $SeS \subseteq I_o = \bigcap_{f \in F} I(f)$ and I_o is closed. Let $I_1 = \{a \in S \mid a^n \in I_o\}$. Then I_1 is closed and $X(e) \subseteq I_1$. Let $a \in I_1$. Then $a^n \mathcal{H} h$ for some $h \in E(S)$ and $a^n \in I_o$. Since $a^n \notin I(h)$, $h \notin F$. So $e|h|a^n$. Thus $a^n \in SeS$ and $a \in X(e)$. So $X(e) = I_1$ is closed.

4 LINEAR ALGEBRAIC GROUPS

Let G be a linear algebraic group. We denote the identity element of G by 1. If $X \subseteq G$, then the normalizer in G of X, $N_G(X) = \{g \in G \mid g^{-1}Xg = X\}$ and the centralizer in G of X, $C_G(X) = \{g \in G \mid gx = xg$ for all $x \in X\}$. The center of G, $C(G) = C_G(G)$. Two subsets X, Y of G are conjugate if $g^{-1}Xg = Y$ for some $g \in G$.

Definition 4.1. Let G be a (linear) algebraic group. The unique irreducible component of G containing 1 will be denoted by G^c. Then $G^c \triangleleft G$, G/G^c is a finite group and $\dim G = \dim G^c$. G is connected if $G^c = G$.

We refer to [34; Sections 7.4, 7.5] for the following

Proposition 4.2. Let G be an algebraic group. Then

(i) If U is a dense open subset of G, then $U^2 = G$.

(ii) If H is a constructible subgroup of G, then $\overline{H} = H$.

(iii) If $H_1,\ldots,H_k$ are closed connected subgroups of G, then the subgroup H of G generated by $H_1,\ldots,H_k$ is closed and connected.

Corollary 4.3. Let $\phi: G \to G'$ be a homomorphism of algebraic groups. Then

(i) $\phi(G)$ is a closed subgroup of G' and the kernel, $\ker \phi$ is a closed subgroup of G.

(ii) $\dim G = \dim \phi(G) + \dim \ker \phi$.

We refer to [34; Chapter IV], [108; Theorems 4.3.3, 5.2.2] for the following theorem.

Theorem 4.4. Let G be an algebraic group, H a closed subgroup of G. Then $G/H = \{aH \mid a \in G\}$ can be made into a quasi-projective variety such that the map γ: $G \to G/H$ given by $\gamma(a) = aH$ is an open morphism and

(i) If Y is any variety, then $1\times\gamma$: $Y \times G \to Y \times G/H$ is open.

(ii) If Y is a variety, ϕ: $G \to Y$ a morphism such that $\phi(ah) = \phi(a)$ for all $a \in G$, $h \in H$, then there exists a unique morphism γ: $G/H \to Y$ such that $\phi = \psi\circ\gamma$.

(iii) If $H \triangleleft G$, then G/H is a linear algebraic group.

Definition 4.5. Let G be a connected group. Then

(i) A maximal closed connected solvable subgroup of G is called a Borel subgroup.

(ii) A closed subgroup P of G containing a Borel subgroup is called a parabolic subgroup. If $P \neq G$ and if there are no proper closed subgroups between P and G, then P is a maximal parabolic subgroup.

(iii) A closed connected subgroup T of G is a torus if $T \cong \mathscr{D}_n^*(K)$ for some $n \in \mathbb{Z}^+$.

Remark 4.6. If $G = GL(n,K)$, then $\mathscr{D}_n^*(K)$ is a maximal torus of G and $\mathscr{T}_n^*(K)$ is a Borel subgroup of G.

The following result is due to A. Borel. See [34; Theorem 21.3].

Theorem 4.7. Let G be a connected algebraic group, B a Borel subgroup of G. Then G/B is a projective variety.

See [34; Corollary 21.3C] or [108; Corollary 7.2.7] for the following.

Corollary 4.8. Let $\phi: G \to G'$ be a surjective homomorphism of connected groups. Let T be a maximal torus of G and B a Borel subgroup of G. Then $\phi(T)$ is a maximal torus of G' and $\phi(B)$ is a Borel subgroup of G'.

Definition 4.9. Let G be an algebraic group, X a variety. Then G acts on X (on the right) if for each $x \in X$, $g \in G$, there is associated an element $x \cdot g \in X$ such that

(i) $(x \cdot g_1) \cdot g_2 = x \cdot g_1 g_2$ for all $x \in X$, $g_1, g_2 \in G$.

(ii) $x \cdot 1 = x$ for all $x \in X$.

(iii) The map, $(x,g) \to x \cdot g$ is a morphism from $X \times G$ into X. If $Y \subseteq X, H \subseteq G$, then $Y \cdot H = \{y \cdot h \mid y \in Y, h \in H\}$.

The power of Theorem 4.7 is exhibited by the following well–known result [110; p. 68].

Corollary 4.10. Let G be a connected group acting on the right on a variety X. Let B be a Borel subgroup of G, Y a closed subset of X such that $Y \cdot B$ is closed in X. Then $Y \cdot G$ is closed in X.

Proof. Let $\gamma : G \to G/B$ be given by $\gamma(a) = aB$ and let $\phi = 1 \times \gamma: X \times G \to X \times G/B$. Then ϕ is open by Theorem 4.4 (i). Let $F = \{(x,g) \mid x \in X, g \in G, x \cdot g \in Y \cdot B\}$. Then F is closed in $X \times G$. So $\phi(F) = \sim\phi(\sim F)$ is closed in $X \times G/B$. Let $p: X \times G/B \to X$ denote the projection onto X. By Theorem 4.7, G/B is a projective variety. So by Theorem 2.20, p is a closed morphism. Hence $Y \cdot G = p(\phi(F))$ is closed in X.

The next result is due to A. Borel. We refer to [34; Chapter VIII] for proofs.

Theorem 4.11. Let G be a connected group. Then

(i) All maximal tori of G are conjugate.

(ii) All Borel subgroups of G are conjugate.

(iii) If B is a Borel subgroup of G, then $N_G(B) = B$ and $G = \bigcup_{x \in G} x^{-1}Bx$.

(iv) If B is a Borel subgroup of G, T a torus in B, then $C_G(T)$ is a connected group having $C_B(T)$ as a Borel subgroup. If T is a maximal torus, then $C_G(T) = C_B(T)$ is a nilpotent group.

Theorem 4.11 implies the following result known as the <u>Lie–Kolchin Theorem</u> (see [34; Theorem 17.6]).

<u>Corollary 4.12</u>. Let G be a closed connected subgroup of GL(n,K). If G is solvable, then it is conjugate to a subgroup of $\mathcal{T}_n^*(K)$. If G is a torus, then it is conjugate to a subgroup of $\mathcal{D}_n^*(K)$.

<u>Definition 4.13</u>. Let G be a closed subgroup of GL(n,K), a ∈ G. Then a is <u>unipotent</u> if the only eigenvalue of a is 1. Let $G_u = \{a \in G \mid a \text{ is unipotent}\}$. G is <u>unipotent</u> if $G = G_u$.

<u>Remark 4.14</u>. (i) The above definition is independent of the particular choice of linear representation of G. In fact if $\phi: G \to G'$ is a homomorphism of algebraic groups, then $\phi(G_u) \subseteq G'_u$. See [34; Theorem 15.3].

(ii) By Theorem 4.11 (iv), any unipotent group is nilpotent.

The next theorem is due to A. Borel. We refer to [34; Theorem 19.3] for a proof.

<u>Theorem 4.15</u>. Let G be a connected solvable group, T a maximal torus of G. Then $U = G_u$ is a closed normal subgroup of G, G = TU and G/U ≅ T. Moreover G is nilpotent if and only if G ≅ T × U.

Corollary 4.16. Let G be a connected group, H a closed connected normal subgroup of G. Let T, B be a maximal torus and a Borel subgroup of G, respectively. Then $T \cap H, B \cap H$ are a maximal torus and a Borel subgroup of H, respectively.

Proof. Let T_o, B_o be a maximal torus and a Borel subgroup of H, respectively. Since $H \triangleleft G$, we may assume without loss of generality that $T_o \subseteq T, B_o \subseteq B$. Let $T_1 = H \cap T$, $B_1 = B \cap T$. Then $T_o \subseteq T_1 \subseteq C_H(T_o)$ and $C_H(T_o)$ is nilpotent by Theorem 4.11 (iv). So by Theorem 4.15, $T_o = T$. Now $B_o \subseteq B_1^c$ and hence $B_o = B_1^c \triangleleft B_1$. So by Theorem 4.11 (iii), $B_o = B_1$.

Definition 4.17. Let G be an algebraic group. A homomorphism $\chi: G \to K^*$ is called a character of G. Let $\mathscr{X}(G)$ denote the group of all characters of G.

Remark 4.18. (i) It is easily seen that $\mathscr{X}(G)$ is linearly independent in the vector space of all K–valued functions on G. See [34; Lemma 16.1].

(ii) Let $T = \mathscr{D}_n^*(K)$, $\chi_1,\dots,\chi_n$ the n projections of T into K^*. Then $\chi_1,\dots,\chi_n$ freely generates $\mathscr{X}(T)$. See [34; Section 16.2].

We refer to [34; Section 16.2], [108; Section 2.5] for the following.

Theorem 4.19. Let T be a torus, $\dim T = n$. Then $\mathscr{X}(T) \cong (\mathbb{Z}^n,+)$. Moreover any closed connected subgroup of T is also a torus.

The structure of unipotent groups is more complicated. Even the proof of the following result is not easy (see [34; Theorem 20.5]).

Theorem 4.20. Let G be a unipotent group, $\dim G = 1$. Then $G \cong (K,+)$.

Definition 4.21. Let G be a connected group, T a maximal torus of G. Then

(i) $W = W(G) = N_G(T)/C_G(T)$ is the Weyl group of G. If $\sigma = xC_G(T) \in W$, $t \in T$, then let $t^\sigma = x^{-1}tx \in T$. Thus W is a subgroup of the automorphism group of T.

(ii) If $\chi \in \mathscr{X}(T)$, $\sigma \in W$, then let $\chi\sigma \in \mathscr{X}(T)$ be given by: $\chi\sigma(t) = \chi(t^\sigma)$ for $t \in T$.

(iii) Let $\mathscr{B} = \mathscr{B}(T)$ denote the set of all Borel subgroups of G containing T. If $B \in \mathscr{B}$, $\sigma = xC_G(T) \in W$, then let $B\sigma = Bx$, $\sigma B = xB$, $B^\sigma = \sigma^{-1}B\sigma = x^{-1}Bx$. Note that by Theorem 4.11 (iv), $C_G(T) \subseteq B$.

We refer to [34; Sections 24, 25] for a proof of the following.

Theorem 4.22. Let G be a connected group, T a maximal torus of G. Then

(i) W is a finite group.

(ii) If $B_1, B_2 \in \mathscr{B}(T)$, then there exists a unique $\sigma \in W$ such that $B_1^\sigma = B_2$.

(iii) $|\mathscr{B}(T)| = |W|$

(iv) $|W| = 1$ if and only if G is solvable.

Definition 4.23. Let G be a connected group, T_o a torus in G. Then T_o is regular if $\mathscr{B}(T_o) = \{B|B$ is a Borel subgroup of G containing $T_o\}$ is finite. Otherwise T_o is singular.

We refer to [34; Proposition 24.2] for the following.

Proposition 4.24. Let G be a connected group, T_o a torus in G. Then T_o is regular if and only if $C_G(T_o)$ is solvable. In such a case $\mathscr{B}(T_o) = \mathscr{B}(T)$ and $C_G(T_o) \subseteq B$ for all $B \in \mathscr{B}(T)$.

Corollary 4.25. Let $\phi: G \to G'$ be a surjective homomorphism of connected groups, $H = (\ker \phi)^c$. Then $|W(G)| = |W(H)| \cdot |W(G')|$.

Proof. By Corollary 4.8 and Theorem 4.22, we are reduced to the case when G' is solvable. Let $T_o = T \cap H$. Then T_o is a maximal torus of H by Corollary 4.16. By Theorem 4.22, it suffices to show that $|\mathscr{B}(T)| = |\mathscr{B}_H(T_o)|$. By Corollary 4.16, $\mathscr{B}_H(T_o) = \{B \cap H | B \in \mathscr{B}_G(T_o)\}$. Let $B_1, B_2 \in \mathscr{B}_G(T_o)$ such that $B_1 \cap H = B_2 \cap H$. By Corollary 4.8, $\phi(B_1) = G'$. Hence $G = B_1(\ker \phi)$. Since G is connected, we see that $G = B_1H$. By Theorem 4.11 (ii), there exists $h \in H$ such that $h^{-1}B_1h = B_2$. Then $h \in N_H(B_1 \cap H) = B_1 \cap H$ by Theorem 4.11 (iii). So $B_1 = B_2$ and $|\mathscr{B}_G(T_o)| = |\mathscr{B}_H(T_o)|$ is finite. By Proposition 4.24, $\mathscr{B}(T) = \mathscr{B}_G(T_o)$, completing the proof.

Definition 4.26. Let G be a connected group. Then

(i) The maximal closed connected normal solvable subgroup of G is called the radical of G and is denoted by rad G. The unipotent group, $\text{rad}_u G = (\text{rad } G)_u$ is called the unipotent radical of G.

(ii) G is reductive if $\text{rad}_u G = \{1\}$. G is semisimple if $\text{rad } G = \{1\}$.

(iii) G is simple if G has no closed connected normal subgroups other than $\{1\}$ and G, and is non abelian.

(iv) If T is a maximal torus of G, then the rank of G, rank G = dim T. The semisimple rank of G, $\text{rank}_{ss} G = \text{rank } (G/\text{rad } G)$.

Remark 4.27. (i) GL(n,K) is a reductive group and SL(n,K) is a simple algebraic group. The direct product of simple algebraic groups is semisimple.

(ii) Let G be a simple algebraic group. Then G need not be simple as an abstract group. However $C = C(G)$ is finite and G/C is simple as an abstract group. See [34; Corollary 29.5].

(iii) Let $\phi: G \to G'$ be a surjective homomorphism of connected groups. Then $\phi(\text{rad } G) = \text{rad } G'$ and $\phi(\text{rad}_u G) = \text{rad}_u G'$. In particular $G/\text{rad } G$ is a semisimple group and $G/\text{rad}_u G$ is a reductive group.

(iv) If G is a connected group, then $\mathrm{rad}\, G$ is just the identity component of the intersection of all Borel subgroups of G. If H is a closed connected normal subgroup of G, then $\mathrm{rad}\, H \triangleleft G$, $\mathrm{rad}_u H \triangleleft G$. Hence $\mathrm{rad}\, H \subseteq \mathrm{rad}\, G$, $\mathrm{rad}_u H \subseteq \mathrm{rad}_u G$.

Definition 4.28. Let G be a group, H_1, H_2 subgroups of G. Then (H_1,H_2) is the subgroup of G generated by $h_1h_2h_1^{-1}h_2^{-1}$ $(h_1 \in H_1, h_2 \in H_2)$.

See [34; Proposition 17.2] for the following.

Proposition 4.29. Let G be an algebraic group, H_1, H_2 closed subgroups of G. Then

(i) If H_1, H_2 are connected, then (H_1, H_2) is a closed connected subgroup of G.

(ii) If H_1 or H_2 is normal, then (H_1, H_2) is a closed normal subgroup of G.

See [34; Theorem 27.5] for the following.

Theorem 4.30. Let G be a semisimple group. Then $G = G_1 ... G_n$ where $G_1,...,G_n$ are the closed normal simple subgroups of G. Moreover,

(i) $G_i = (G_i,G_i)$, $G = (G,G)$.

(ii) $(G_i,G_j) = \{1\}$ for $i \neq j$ and the product map from $G_1 \times ... \times G_n$ onto G has a finite kernel.

(iii) If H is a closed connected normal subgroup of G, then $H = G_{i_1} ... G_{i_m}$ for some subset $\{i_1,...,i_m\}$ of $\{1,...,n\}$.

Remark 4.31. Let G be a closed connected subgroup of $GL(n,K)$. Then $G/\mathrm{rad}\, G$ is a semisimple group. Hence by Theorem 4.30 (i), $G = (G,G)\mathrm{rad}\, G$. In particular if $\mathrm{rad}\, G$ is unipotent, then $\det a = 1$ for all $a \in G$.

By Theorem 4.30, [108; Proposition 6.15], we have,

Theorem 4.32. Let G be a reductive group. Then $C(G)$ is the intersection of all Borel subgroups of G, rad $G = C(G)^c$ is a torus, (G,G) is a semisimple group and $G = (G,G)\text{rad } G$.

Remark 4.33. Let G be a connected group. Then by Remark 4.27 (iii), Theorem 4.32, $(G, \text{rad } G) \subseteq \text{rad}_u G$.

Corollary 4.34. Let G be a reductive group, $C = C(G)$, $G' = (G,G)$. Let H be a closed normal subgroup of G. Then $H = C'H'$ where $C' = C \cap H$ and $H' = (H^c,H^c) \triangleleft G'$.

Proof. Let $G_1,\dots,G_n$ be the simple components of G'. Now for each $i = 1,\dots,n$, $(H,G_i) \subseteq H \cap G_i \triangleleft G_i$. Hence either $H \cap G_i = G_i$ or else $H \cap G_i \subseteq C$. Let $h \in H$. Then $h = cg_1 \dots g_n$ for some $c \in C$, $g_i \in G_i$, $i = 1,\dots,n$. Suppose $G_i \not\subseteq H$. Then for all $g \in G_i$, $g_i g g_i^{-1} g^{-1} = hgh^{-1}g^{-1} \in (H,G_i) \subseteq C$. By Remark 4.27 (ii), $g_i \in C$. Hence $H \subseteq CH'$ and the result follows.

The next result is known as the Bruhat decomposition. See [31; Theorem 28.3].

Theorem 4.35. Let G be a reductive group, T a maximal torus of G. Then for any $B, B' \in \mathscr{B}(T)$, G is the disjoint union of $B\sigma B'$ $(\sigma \in W)$.

See [34; Corollary 28.3] for the following.

Corollary 4.36. Let G be a reductive group, B, B' Borel subgroups of G. Then $B \cap B'$ contains a maximal torus of G.

Definition 4.37. Let G be a reductive group, B, B' Borel subgroups of G. Then B, B' are opposite if $B \cap B'$ is a torus. In such a case, B' is the opposite Borel subgroup of G relative to $T = B \cap B'$.

See [34; Section 26.2] for the following.

Theorem 4.38. Let G be a reductive group, T a maximal torus of G. Then every $B \in \mathcal{B}(T)$ has a unique opposite $B^- \in \mathcal{B}(T)$ relative to T. In particular $C_G(T_o)$ is a reductive group for any torus $T_o \subseteq T$ and $C_G(T) = T$.

Remark 4.39. (i) If B, B^- are opposite relative to T, then $B, b^{-1}B^-b$ are opposite relative to $b^{-1}Tb$ for any $b \in B$.

(ii) If G is a connected group, then $G/\mathrm{rad}_u G$ is a reductive group. Hence for any maximal torus T of G, $\cap\, \mathcal{B}(T) = T\, \mathrm{rad}_u G$.

(iii) Let $G = GL(n,K)$, $T = \mathcal{D}_n^*(K)$, $B = \mathcal{T}_n^*(K)$, $B^- = B^t$. Then B, B^- are opposite relative to T.

Definition 4.40. Let G be a reductive group, P, P' parabolic subgroups of G. Then P, P' are opposite if $P \cap P'$ is a reductive group. In such a case, if $T \subseteq P \cap P'$ is a maximal torus, then we say that P' is opposite to P relative to T.

The following theorem is due to Borel and Tits [5].

Theorem 4.41. Let G be a reductive group, T a maximal torus of G, P a parabolic subgroup of G containing T. Then P has a unique opposite P^- relative to T. Moreover $P = LU$, $P^- = LU^-$ where $L = P \cap P^-$, $U = \mathrm{rad}_u P$, $U^- = \mathrm{rad}_u P^-$.

In the above theorem, L is called a Levi factor of P. For the rest of this chapter, fix a reductive group G, a maximal torus T of G and a Borel subgroup B containing T. Let B^- denote the opposite of B, relative to T, $\dim T = m$. Then $\mathcal{X}(T) \cong (\mathbb{Z}^m,+) \subseteq (\mathbb{R}^m,+)$. We will view $\mathcal{X}(T)$ additively.

Definition 4.42. Let $\lambda: G \to GL(V)$ be a finite dimensional representation. Then $\chi \in \mathscr{X}(T)$ is a weight of λ if $V_\chi = \{v \in V | \lambda(t)v = \chi(t)v$ for all $t \in T\} \neq \{0\}$. In such a case V_χ is called the weight space of χ. Then since $\lambda(T)$ is diagonalizable, $V = V_{\chi_1} \oplus .. \oplus V_{\chi_k}$.

As in the case of Lie groups, the 'tangent space' of G at 1 forms a Lie algebra $\mathscr{L} = \mathscr{L}(G)$. Moreover, $\dim \mathscr{L} = \dim G$. Also, G acts as a group of automorphisms of $\mathscr{L}$. This gives rise to the adjoint representation, $\mathrm{Ad}: G \to GL(\mathscr{L})$. The kernel of this representation is just the center of G. We refer to [34; Chapter III], [108; Chapter 3] for details. The basic example to keep in mind is $G = GL(n,K)$ in which case $\mathscr{L}(G) = \mathscr{M}_n(K)$ with $[x,y] = xy - yx$. Also $\mathrm{Ad}(g)(a) = gag^{-1}$ for $g \in G$, $a \in \mathscr{M}_n(K)$.

Definition 4.43. The non–zero weights (in the additive notation) of $\mathrm{Ad}: G \to GL(\mathscr{L}(G))$ are called the roots of G and denoted by ϕ. If $\alpha \in \phi$, let $T_\alpha = (\ker \alpha)^c$, $G_\alpha = C_G(T_\alpha)$, $\mathscr{L}_\alpha$ the weight space of α.

Example 4.44. Let $G = GL(3,K)$. Then $\phi = \{\chi_1,\chi_2,\chi_3, -\chi_1,-\chi_2,-\chi_3\}$, where χ_1: $\mathrm{diag}(a,b,c) \to a/b$, χ_2: $\mathrm{diag}(a,b,c) \to b/c$, χ_3:$\mathrm{diag}(a,b,c) \to a/c$, $-\chi_1$: $\mathrm{diag}(a,b,c) \to b/a$, $-\chi_2$: $\mathrm{diag}(a,b,c) \to c/b$, $-\chi_3$: $\mathrm{diag}(a,b,c) \to c/a$. Note that $\Delta = \{\chi_1,\chi_2\}$ forms a basis for the space spanned by ϕ, and $\chi_3 = \chi_1 + \chi_2$.

See [34; Chapter IX] for the following.

Theorem 4.45. (i) $\phi = -\phi$, $\phi W = \phi$.

(ii) $T_\alpha (\alpha \in \phi)$ are exactly the maximal singular tori of G contained in T.

(iii) If $\alpha \in \phi$, then $W(G_\alpha) = \{1, \sigma_\alpha\}$, $\phi(G_\alpha) = \{\alpha,-\alpha\}$, $\alpha\sigma_\alpha = -\alpha$.

(iv) If $\alpha \in \phi$, then $\mathscr{L}(\mathrm{rad}_u(B \cap G_\alpha))$ is either $\mathscr{L}_\alpha$ or $\mathscr{L}_{-\alpha}$.

Definition 4.46. (i) If $\alpha \in \phi$, then $\sigma_\alpha: \alpha \to -\alpha$ is called a reflection. We let $T_{\sigma_\alpha} = T_\alpha = \{t \in T | t^{\sigma_\alpha} = t\}^c$.

(ii) If $\alpha \in \phi$, then $U_\alpha = \mathrm{rad}_u(B \cap G_\alpha)$ if $\mathcal{L}(\mathrm{rad}_u(B \cap G_\alpha)) = \mathcal{L}_\alpha$. Otherwise $U_\alpha = \mathrm{rad}_u(B^- \cap G_\alpha)$. Thus $\mathcal{L}(U_\alpha) = \mathcal{L}_\alpha$ and $T \subseteq N_G(U_\alpha)$. U_α's are called root subgroups.

See [34; Theorem 26.3] for the following.

Theorem 4.47. Let $\alpha \in \phi$. Then

(i) $\dim U_\alpha = 1$ and there is an isomorphism $\varepsilon_\alpha: (K,+) \to U_\alpha$ such that for all $t \in T, x \in K$, $t\varepsilon_\alpha(x)t^{-1} = \varepsilon_\alpha(\alpha(t) \cdot x)$.

(ii) For all $\sigma \in W$, $\sigma^{-1}U_\alpha\sigma = U_{\alpha\sigma}$.

(iii) G is generated by $U_\alpha(\alpha \in \phi)$ and T.

Remark 4.48. $\alpha \in \phi$ is positive (relative to B) if $U_\alpha \subseteq B$. Let $\phi^+ = \{\alpha \in \phi | \alpha$ is positive relative to $B.\}$ Then $\phi^- = -\phi^+$ is the set of positive roots relative to B^-. Moreover ϕ is the disjoint union of ϕ^+ and ϕ^-. Let $\Delta = \Delta(B) = -\Delta(B^-) = \{\alpha \in \phi^+ | \alpha$ is not a non–negative linear combination of $\phi^+ \backslash \{\alpha\}\}$. Then Δ is called the base of ϕ, relative to B. It turns out that Δ is a basis (over $\mathbb{R}$) for the span of ϕ and every element of ϕ^+ is a non–negative integral linear combination of Δ. $\sigma_\alpha(\alpha \in \Delta)$ are called the simple reflections relative to B. Let $\mathcal{S} = \mathcal{S}(B) = \mathcal{S}(B^-) = \{\sigma_\alpha | \alpha \in \Delta\}$. Then $|\phi^+| = |\mathcal{S}| = \mathrm{rank}_{ss} G$ and W is generated by $\mathcal{S}$. The map: $B \to \Delta(B)$ is injective and for $\sigma \in W$, $\Delta(B^\sigma) = \Delta(B)\sigma$. Moreover $\Delta(B)W = \phi$. Also B_u is generated by $U_\alpha(\alpha \in \phi^+)$. We refer to [108; Chapter 10] for details.

See [34; Proposition 27.2], [110; p. 80] for the following.

Proposition 4.49. Let $\phi: G \to GL(V)$ be a finite dimensional representation such that $\ker \phi \subseteq C(G)$. Let $\alpha \in \phi, \chi \in \mathcal{X}(T)$. Then for all $u \in U_\alpha$, $v \in V_\chi$, $\phi(u)(v) - v$ lies in the sum of $V_{\chi+k\alpha}(k \in \mathbb{Z}^+)$.

The following consequence was pointed out to the author by J. E. Humphreys.

Corollary 4.50. Suppose $G \subseteq GL(n,K)$. Then there exists $a \in GL(n,K)$ such that $a^{-1}Ba$, $a^{-1}B^{-}a$ consist of upper and lower triangular matrices, respectively.

Proof. Let $G \subseteq GL(V)$ and let X denote the set of weights of T. If $\chi_1, \chi_2 \in \mathscr{X}$, define $\chi_1 \leq \chi_2$ if $\chi_1 - \chi_2$ is a non-negative linear combination of ϕ^+. Find bases of $V_\chi (\chi \in \mathscr{X})$ and order them in such a way that if $\chi_1 < \chi_2$, then a basis vector $v_1 \in V_{\chi_1}$ occurs before any basis vector $v_2 \in V_{\chi_2}$. By Proposition 4.49, U_α is upper triangular with respect to this basis for all $\alpha \in \phi^+$ and lower triangular for all $\alpha \in \phi^-$. Moreover T is diagonal. Since B is generated by $T, U_\alpha (\alpha \in \phi^+)$ and B^- is generated by $T, U_\alpha (\alpha \in \phi^-)$, the result follows.

The next result is due to Borel, Tits [5].

Theorem 4.51. If $I \subseteq \mathscr{S}$, let $W_I = \langle I \rangle$, $P_I = BW_IB$, $P_I^- = B^-W_IB^-$. Then

(i) P_I, P_I^- are opposite parabolic subgroups of G relative to T and $W(P_I) = W(P_I^-) = W_I$.

(ii) If P is a parabolic subgroup of G containing B, then $P = P_I$ for some $I \subseteq \mathscr{S}$.

(iii) If $I, I' \subseteq \mathscr{S}$, $x \in G$, $x^{-1}P_Ix \subseteq P_{I'}$, then $x \in P_{I'}$ and $I \subseteq I'$.

Definition 4.52. Two parabolic subgroups P, P' of G are of the same type if they are conjugate. P, P' are of opposite type if P' is conjugate to an opposite of P.

Corollary 4.53. Let $\alpha \in \Delta(B)$, $U = B_u$. Let $P = B \cup B\sigma_\alpha B$, $Y = \mathrm{rad}_u(P)$. Then $\sigma_\alpha Y = Y\sigma_\alpha$ and $YU_\alpha = U_\alpha Y = U$.

Proof. By Theorem 4.51, $W(P) = \{1,\sigma_\alpha\}$. Since $Y \triangleleft P$, $\sigma_\alpha Y = Y\sigma_\alpha$. Now $\sigma_\alpha U_\alpha \sigma_\alpha^{-1} = U_{-\alpha} \not\subset B$. Hence $U_\alpha \not\subset Y$. By Theorem 4.45 (i), $\dim U/Y = 1$. It follows that $U_\alpha Y = YU_\alpha = U$.

Renner [96; Proposition 7.4] derives the following result from the classification of reductive groups [108; Theorem 11.4.3].

Proposition 4.54. G admits an involution $*$ such that $t^* = t$ for all $t \in T$ and $U_\alpha^* = U_{-\alpha}$ for all $\alpha \in \phi$.

Remark 4.55. (i) In the above situation, it is clear that $P_I^* = P_I^-$. Thus for any parabolic subgroup P of G, P, P^* are of opposite type.

(ii) If H is a closed normal subgroup of G, then it follows from Corollary 4.34 that $H^* = H$.

Now assume that (G,G) is simple and let $\Delta = \Delta(B)$, $\mathscr{S} = \mathscr{S}(B)$. If $\alpha,\gamma \in \Delta$, then $\gamma - \gamma\sigma_\alpha$ turns out to be an integral multiple of α. This integer is denoted by $< \gamma,\alpha >$ and called a Cartan integer. The matrix of Cartan integers gives rise to the various possibilities for the root systems: $A_l (l \geq 1)$, $B_l (l \geq 2)$, $C_l (l \geq 3)$, $D_l (l \geq 4)$, E_6, E_7, E_8, F_4, G_2. See [34], [108]. The Weyl group $W = <\mathscr{S}>$ is a special type of a finite group, called a Coxeter group. If $\alpha, \gamma \in \Delta$, let $m(\alpha,\gamma)$ denote the order of $\sigma_\alpha\sigma_\gamma$ Then W is completely determined by the relations $(\sigma_\alpha\sigma_\gamma)^{m(\alpha,\gamma)} = 1$. It turns out that for $\alpha \neq \gamma$, $m(\alpha,\gamma) = 2,3,4$ or 6. If $m(\alpha,\gamma) = 3$, define σ_α ——— σ_β, if $m(\alpha,\gamma) = 4$ define σ_α ═══ σ_γ, if $m(\alpha,\gamma) = 6$, define σ_α ≡≡≡ σ_γ. The possibilities are then given by the following diagrams [12], [115].

A_l: •——• · · · •——•

B_l or C_l: •——• · · · •——•══•

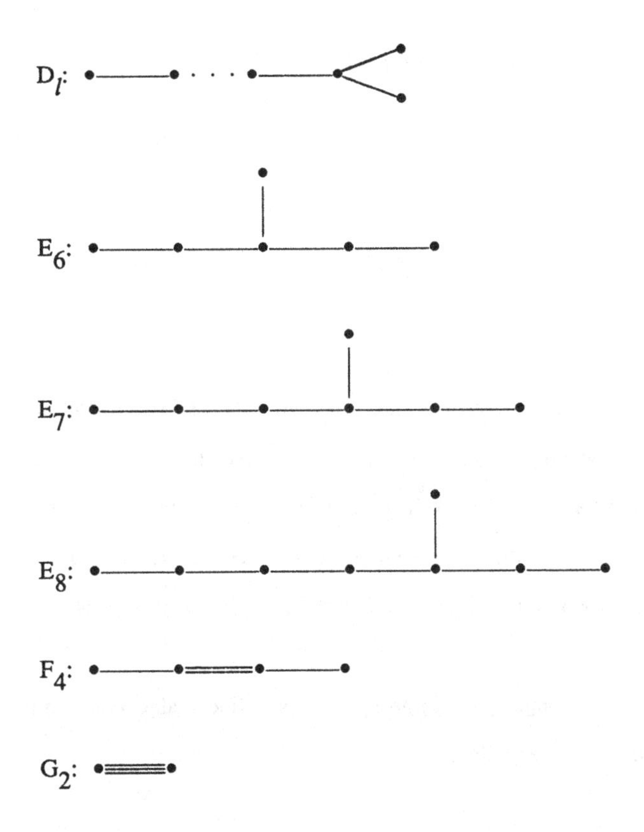

Thus the Weyl group does not distinguish between types B_l, C_l. The Cartan matrix can be described completely via the <u>Dynkin diagrams</u> which contain slightly more information than the above (Coxeter) diagrams (see [34; Appendix]). When (G,G) is not simple, its diagram is <u>reducible</u> in that it is the disjoint union of the diagrams of the simple components.

For more details on algebraic groups, we refer to Borel [4], Carter [6], Hochschild [32], Humphreys [34], Springer [108] and Steinberg [110].

5 CONNECTED ALGEBRAIC SEMIGROUPS

For algebraic groups, the topological terms 'irreducible' and 'connected' have the same meaning. For algebraic semigroups, this is not so. Consider, for example, $S = \{\text{diag}(a,b) \mid a^2 = b^2\} \subseteq \mathscr{D}_2(K)$. Topologically connected algebraic monoids are briefly studied by Renner [101]. However, we will use the term 'connected semigroup' to mean that the underlying variety is irreducible.

Definition 5.1. A connected semigroup S is a linear algebraic semigroup whose underlying variety is irreducible.

Remark 5.2. Let M be a linear algebraic monoid and let M_1, M_2 be irreducible components of M containing 1. Then the product map from $M_1 \times M_2$ into M shows that $\overline{M_1 M_2}$ is irreducible. Clearly $M_1, M_2 \subseteq \overline{M_1 M_2}$. Hence $M_1 = M_2$. Thus 1 lies in a unique irreducible component M^c of M. Clearly M^c is a connected monoid.

Remark 5.3. Let S be a connected algebraic semigroup, $e \in E(S)$. Then eS, Se, eSe, $\overline{SeS}$ are connected semigroups.

Remark 5.4. Let $\phi: G_0 \to GL(n,K)$ be a representation of a connected group G_o. Then $M = M(\phi) = \overline{K\phi(G_o)} \subseteq \mathscr{M}_n(K)$ is a connected monoid with zero.

Example 5.5. Let M be a connected algebraic monoid with group of units G. Let S' be an irreducible component of $S = M\backslash G$. Then S' is a connected semigroup which is an ideal of M.

Example 5.6. Let $M = K^4$ with multiplication

$$(a,b,c,d)(a',b',c',d') = (aa',ab' + bd', dc' + ca', dd').$$

Then M is a connected monoid with zero. Let $e = (1,0,0,0) \in E(M)$. Then MeM is not closed. See [65; Example 4.1].

Example 5.7. Let $S = K^3$ with

$$(a,b,c)(a',b',c') = (aa' + aba'c',b',c).$$

Then S is a connected regular semigroup. Let $J_1 = \{(a,b,c) \in S \mid a \neq 0\}$, $J_2 = \{(a,b,c) \in S \mid a = 0\}$. Then $\mathcal{U}(S) = \{J_1,J_2\}$, $J_1 > J_2$. Let $f = (0,1,-1) \in E(J_2)$. Then there is no $e \in E(J_1)$ with $e \geq f$. See [65; Example 4.11]. Contrast this situation with Corollary 6.9.

The following result is from the author [67; Theorem 8].

Proposition 5.8. Let S be a linear algebraic semigroup, $e \in E(S)$, J,R,L the $\mathcal{J}$-class, $\mathcal{R}$-class, $\mathcal{L}$-class of e, respectively. Then $E(J)$, $E(R)$, $E(L)$ are closed subsets of S. If S is a connected semigroup, these sets are also irreducible.

Proof. By Corollary 3.16, we may assume that S is a closed subsemigroup of some $\mathcal{M}_n(K)$. Let $\rho(e) = k$. If $a \in S$, let $\delta(a)$ = the sum of products of k eigenvalues of a. Since $\delta(a)$ is a co-efficient of the characteristic polynomial of a, $\delta: S \to K$ is a morphism. Let $X = \{f \mid f \in \overline{SeS}, f^2 = f, \delta(f) = k\}$. Then X is closed, $E(J) \subseteq X$. If $f \in X$, then by Corollary 3.30, $f \in SeS$. So $\rho(f) \leq k$. Since $\delta(a) = k$, we see that $\rho(f) = k$. There exist $x,y \in S$ such that $xey = f$. Let $f' = eyfxe \in E(S)$. Then $e \geq f'$. Since $f \mathcal{J} f'$, $\rho(f') = k$. So $e = f' \mathcal{J} f$ in S. Hence $E(J) = X$ is closed. Clearly $E(R) = \{f \mid f \in E(S), ef = f, fe = e\}$, $E(L) = \{f \mid f \in E(S), fe = f, ef = e\}$ are closed sets.

Now assume that S is a connected semigroup, $U = \{(x,y) \mid x,y \in S, \det_e(yx) \neq 0\}$. Then U is a non-empty open (hence irreducible) subset of $S \times S$. Then $(x,y) \to (eyxe)^{-1}$ is a morphism on U where the inverse is taken in the $\mathcal{H}$-class of e. Consider the morphism $\phi: U \to E(J)$ given by $\phi(x,y) = x(eyxe)^{-1}y$. Let $f \in E(J)$. Then $xey = f$ for some $x,y \in S$. So $eyfxe \in E(J)$ and $e = eyfxe$. Thus $(x,yf) \in U$, $\phi(x,yf) = f$. Hence $\phi(U) = E(J)$ is irreducible. Let $V = \{a \mid a \in eS, \det_e(a) \neq 0\}$. Define a morphism $\psi: V \to E(R)$ as $\psi(a) = (eae)^{-1}a$. If $f \in E(R)$, then $\psi(f) = f$. Hence $\psi(V) = E(R)$ is irreducible. Similarly $E(L)$ is irreducible.

The following result of the author [64; Theorem 2.16] turns out to be quite useful.

Theorem 5.9. Let S be a connected semigroup, $e,f \in E(S)$, $e \mathcal{J} f$. Then there exist $e_1,e_2,f_1,f_2 \in E(S)$ such that $e \mathcal{R} e_1 \mathcal{L} f_1 \mathcal{R} f$ and $e \mathcal{L} e_2 \mathcal{R} f_2 \mathcal{L} f$.

Proof. Let $e,f \in E(S)$, $e \mathcal{J} f$. We claim that there exists $e_1 \in E(S)$ such that $e \mathcal{R} e_1$, $e_1 f \mathcal{J} f$. Suppose not. Let H, H' denote the $\mathcal{H}$-class of e,f, respectively. By Remark 3.23, $eSe\backslash H$, $fSf\backslash H'$ are closed sets. There exist $x,y \in S$ such that $xey = f$. We have the following closed subsets of eS:

$$X = \{a \in eS \mid fxaf \in fSf\backslash H'\}$$

$$Y = \{a \in S \mid ae \in eSe\backslash H\}.$$

Suppose $e \notin X$. Then $fxef \in H'$ and $ef|f$. Hence $ef \mathscr{J} f$, a contradiction. So $e \in X$. Clearly $fxeyf = f$, whereby $ey \notin X$. Also $e \notin Y$. We claim that $ef \in Y$. Otherwise $efe \in H$ and $ef|e|f$, a contradiction. Hence $ef \in Y$. Since eS is irreducible, we see that $eS \neq X \cup Y$. So there exists $a \in eS$ such that $a \notin X \cup Y$. Then $ea = a$, $fxaf \mathscr{H} f$, $ae \mathscr{H} e$. So there exists $z \in S$ such that $zae = e$. Then $za^2 = zaea = ea = a$. Hence $a^2 \mathscr{J} a$. By Theorem 1.4 (v), there exists $e_1 \in E(S)$ such that $a \mathscr{H} e_1$. Then $e \mathscr{R} e_1$. Also $e_1f|af|fxaf|f$. Hence $e_1f \mathscr{J} f$, a contradiction. Thus there exists $e_1 \in E(S)$ such that $e \mathscr{R} e_1$, $e_1f \mathscr{J} f$. By Theorem 1.4 (vi), there exists $f_1 \in E(S)$ such that $e_1 \mathscr{L} f_1 \mathscr{R} f$. Similarly there exists $e_2,f_2 \in E(S)$ such that $e \mathscr{L} e_2 \mathscr{R} f_2 \mathscr{L} f$. This proves the theorem.

The following result is due to the author [65; Theorem 2.7].

Theorem 5.10. Let S be a connected semigroup. Then $\mathscr{U}(S)$ is a finite lattice.

Proof. We can assume by Corollary 3.16 that S is a closed subsemigroup of some $\mathscr{M}_n(K)$. Let $E = E(S)$. If $e \in E$, let $I(e) = \{a \in S \mid a^n \in SeS\}$. Then $I(e)$ is a closed subset of S by Corollary 3.30. Since $\mathscr{U}(S)$ is finite by Theorem 3.28, the family $\{SeS \mid e \in E\}$ and hence the family $\mathscr{A} = \{I(e) \mid e \in E\}$ is finite. By Theorem 3.18, $S = \bigcup_{e \in E} I(e)$. Since S is a connected semigroup, $I(v) = S$ for some $v \in E$. Then clearly J_v is the maximum element of $\mathscr{U}(S)$. Since $\mathscr{U}(S)$ is finite, it suffices to show that $\mathscr{U}(S)$ is a $\wedge$-semilattice. So let $e,f \in E$ and let $\Gamma = \{g \mid g \in E, e|g, f|g\}$. Let $I = \bigcup_{g \in \Gamma} I(g)$. Let $x,y,z \in S$. Then $(xeyfz)^n \mathscr{H} g$ for some $g \in \Gamma$. So $g \in \Gamma$ and $xeyfz \in I(g)$. Define $\phi: S \times S \times S \to I$ as $\phi(x,y,z) = xeyfz$. Since $S \times S \times S$ is irreducible and $\mathscr{A}$ is finite, we see that $\phi(S \times S \times S) \subseteq I(h)$ for some $h \in \Gamma$. Then clearly $J_e \geq J_h$, $J_f \geq J_h$. Let $J \in \mathscr{U}(S)$ such that $J_e \geq J$, $J_f \geq J$. Let $g \in E(J)$. Then $e|g$, $f|g$. So $xey = g = sft$ for some $x,y,s,t \in S$. So $g = \phi(x,y,s,t) \in I(h)$. Thus $h|g$ and $J_h \geq J$. Hence $J_e \wedge J_f = J$. This proves the theorem.

Remark 5.11. The open problem then is to determine all possible $\mathcal{U}(S)$. If e is an idempotent in the maximum $\mathcal{J}$-class of S, then $\mathcal{U}(S) \cong \mathcal{U}(M)$ where $M = eSe$. In fact, there exists a connected regular monoid M' with zero such that $\mathcal{U}(M) \cong \mathcal{U}(M')$ (see the proof of Theorem 15.1). When $\mathcal{U}(M')\backslash\{0\}$ has a minimum element, the possibilities are determined in [89].

The next result is taken from the author [73; Theorem 2.1].

Theorem 5.12. Let S be a connected semigroup with zero 0. Then the following conditions are equivalent.

(i) S is completely regular.

(ii) S has no non-zero nilpotent elements.

(iii) S is a monoid and the group of units of S is a torus.

(iv) S is isomorphic to a closed submonoid of some $\mathcal{D}_n(K)$ with 0 being the zero matrix.

Proof. That (iii) => (iv) follows from Theorem 3.15, Corollary 4.12. That (iv) => (i) follows from Corollary 3.19. That (i) => (ii) is obvious. So we are left with showing that (ii) => (iii). Thus assume that S has no non-zero nilpotent elements. By Corollary 3.16 we can assume that S is a closed subsemigroup of some $\mathcal{M}_n(K)$. Hence a^n lies in a subgroup of S for all $a \in S$. Let $e \in E(S)$. Suppose $Se \neq eSe$ and consider the morphism $\phi: Se \to eSe$ given by $\phi(x) = ex$. By Theorem 2.21, $\dim \phi^{-1}(0) > 0$. So there exists $x \in Se$, $x \neq 0$ such that $ex = 0$. Then $x^2 = xex = 0$, a contradiction. Hence $Se = eSe$. Similarly $eS = eSe$. Thus the idempotents of S lie in the center of S. By Theorem 5.10, $\mathcal{U}(S)$ has a maximum element J. Let $E(J) = \{h\}$. Then for all $a \in S$, $a^n \in ShS = hS$. Suppose $S \neq hS$ and consider the morphism $\psi: S \to hS$ given by $\psi(a) = ha$. Then by Theorem 2.21, $\dim \psi^{-1}(0) > 0$. So there exists $a \in S$, $a \neq 0$ such that $ha = 0$. Then $a^n = ha^n = 0$, a contradiction. Hence $S = hS$ and $h = 1$ is the identity element of S. We may assume that 1 is the identity matrix. We see by Corollary 1.6 that for all $a,b \in S$, $a|b$ implies $a^n|b^n$. Now let $a \in S$. Then $a^n \mathscr{H} e$ for some $e \in E(S)$. Let $S_1 = \overline{aS} \supseteq eS$. Then for all $x \in S$, $e|a^n|(ax)^n$. So $u^n \in eS$ for all $u \in S_1$. Suppose $S_1 \neq eS$ and consider the surjective morphism $\theta: S_1 \to eS$ given by $\theta(x) = ex$. By Theorem 2.21, $\theta^{-1}(0) \neq \{0\}$. So there exists $b \in S_1$ such that $eb = 0$, $b \neq 0$. Then $b^n = eb^n = 0$, a contradiction. Hence $S_1 = eS$ and $a \in eS$. So $a \mathscr{H} e$ and S is a semilattice of groups by Theorem 1.12.

Let G denote the group of units of S. We prove by induction on $\dim S$ that G is a torus. Let $e \in E(S)$ such that 1 covers e. Consider the homomorphism $\gamma: S \to eS$ given by $\gamma(x) = ex$. By the induction hypothesis applied to eS, $eG = \gamma(G)$ is a torus. Let $S_e = \gamma^{-1}(e)^c$. Let V be an irreducible component of $\gamma^{-1}(e)$ containing e. By Theorem 2.21, $V \neq \{e\}$. So there exists $v \in V$, $v \neq e$ such that $ev = e$. Then $v \in G$. So $e \in Vv^{-1} \subseteq S_2$. Let $G_e = G \cap S_e$. Consider the homomorphism $\det: S_e \to K$. Since $\det^{-1}(0) = \{e\}$, we see by Theorem 2.21 that $\dim S_e = 1$. If G_e is unipotent, then $\det(S_e) = \{1,0\}$, a contradiction since S_e is a connected monoid. So by Theorem 4.11 (iv), G_e is a torus. Let T be a maximal torus of G containing G_e. By Corollary 4.8, $\gamma(T) = \gamma(G)$. By Corollary 4.3, $G = T$, completing the proof.

Most of this book has to do with connected regular monoids M with zero. However, the following is clearly an important open problem.

<u>Problem 5.13</u>. Study connected regular semigroups with zero.

Let S be a connected regular semigroup with zero, e an idempotent in the maximum $\mathscr{J}$-class. Then $M = eSe$ is a connected regular monoid with zero. The problem then is to determine the possible S for a given M. A good starting point would be to take $M = \mathscr{M}_n(K)$.

6 CONNECTED ALGEBRAIC MONOIDS

In this chapter we develop the machinery for studying connected monoids. If M is any linear algebraic monoid, then the identity component M^c of M is a connected monoid (see Remark 5.2). If G is the group of units of M, then G is an open subset of M (see Remark 3.23), $M^c = \overline{G^c}$. If G_o is a connected group, $\phi: G_o \to GL(n,K)$ a representation of G_o, then $M = \overline{K\phi(G_o)}$ is a connected monoid with zero. More generally Renner [91] and Waterhouse [116] have shown that any connected group with a non–trivial character occurs as the group of units of a connected monoid with zero.

Let M be a connected monoid with group of units G. Let $X,Y \subseteq M$. Then X,Y are <u>conjugate</u> if there exists $g \in G$ such that $g^{-1}Xg = Y$. The <u>centralizer</u> <u>of</u> X <u>in</u> Y, $C_Y(X) = \{y \in Y | xy = yx$ for all $x \in X\}$ and the <u>normalizer</u> <u>of</u> X <u>in</u> Y, $N_Y(X) = \{y \in Y | Xy = yX\}$. If $\Gamma \subseteq E(M)$, then the <u>right</u> <u>centralizer</u> <u>of</u> Γ <u>in</u> X, $C^r_X(\Gamma) = \{x \in X | xe = exe$ for all $e \in \Gamma\}$, the <u>left</u> <u>centralizer</u> <u>of</u> Γ <u>in</u> X, $C^\ell_X(\Gamma) = \{x \in X | ex = exe$ for all $e \in \Gamma\}$. The <u>center</u> of M, $C(M) = \{x \in M | xy = yx$ for all $y \in M\}$. For $e \in E(M)$, let

$$M_e = \{a \in M | ae = ea = e\}^c, G_e = G \cap M_e.$$

Let T be a maximal torus of G. Then $N_G(T) = N_{\overline{G}}(T)$ and $C_G(T) = C_{\overline{G}}(T)$. If $\sigma = xC_G(T) \in W(G)$, $a \in \overline{T}$, then let $a^\sigma = x^{-1}ax \in \overline{T}$. We will also denote $W(G)$ by $W(M)$.

The following result is due to the author [67], [83].

Proposition 6.1. Let M be a connected monoid with group of units G and let $a,b \in M$. Then

(i) $a \mathscr{J} b$ if and only if $\overline{MaM} = \overline{MbM}$ if and only if $b \in GaG$.

(ii) $a \mathscr{R} b$ if and only if $\overline{aM} = \overline{bM}$ if and only if $b \in aG$.

(iii) $a \mathscr{L} b$ if and only if $\overline{Ma} = \overline{Mb}$ if and only if $b \in G\,a$.

Proof. Define $\phi: G \times G \to \overline{MaM}$ as $\phi(g_1,g_2) = g_1 a g_2$. Then ϕ is a dominant morphism and $\overline{MaM}$ is irreducible. So by Theorem 2.19, there exists a non-empty open subset U of $\overline{MaM}$ such that $U \subseteq GaG$. Similarly there exists a non-empty open subset V of $\overline{MbM}$ such that $V \subseteq GbG$. So if $\overline{MaM} = \overline{MbM}$, then $\emptyset \neq U \cap V \subseteq GaG \cap GbG$. This proves (i). (ii), (iii) are proved similarly.

Proposition 6.2. Let M be a connected monoid with group of units G, $\dim M = p$, $M \neq G$. Let $S_1,\dots,S_k$ denote the irreducible components of $S = M \backslash G$. Then S_i is an ideal of M and $\dim S_i = p-1$, $i = 1,\dots,k$.

Proof. By Theorem 3.15 we can assume that M is a closed submonoid of some $\mathscr{M}_n(K)$. Consider $\phi: M \to K$ given by $\phi(a) = \det a$. Clearly $S = \phi^{-1}(0)$. By Theorem 2.21, $\dim S_i = p - 1$, $i = 1,\dots,k$. Now $S_i \subseteq \overline{MS_iM} \subseteq S$; and $\overline{MS_iM}$ is irreducible, being the closure of the product map from $M \times S_i \times M$ into S. So $S_i = \overline{MS_iM}$ and the result follows.

The following consequence of Corollary 4.10 has been noted in [66], [91]. We follow [91].

Proposition 6.3. Let M be a connected monoid with group of units G and let B be a Borel subgroup of G. Then $M = \overline{B}G = G\overline{B} = \bigcup_{x \in G} x^{-1}\overline{B}x$.

Proof. G acts on M in three ways: $a \cdot g = ag$, $a \cdot g = g^{-1}a$, $a \cdot g = g^{-1}ag$ where $g \in G$, $a \in M$. In each case $\overline{B} \cdot B = \overline{B}$ and $G \subseteq \overline{B} \cdot G$. By Corollary 4.10, $\overline{B} \cdot G$ is closed in M. It follows that $\overline{B} \cdot G = M$.

Corollary 6.4. Let M be a connected monoid with zero 0 and group of units G. Then $0 \in \overline{T}$ for any maximal torus T of G.

Proof. By Proposition 6.3, we can assume that G is solvable. By Remark 3.17, we can assume that M is a closed submonoid of some $\mathcal{M}_n(K)$ with 0 being the zero matrix. By Theorem 4.11, we can further assume that $G \subseteq \mathcal{T}_n^*(K)$, $T \subseteq \mathcal{D}_n^*(K)$. Then $M \subseteq \mathcal{T}_n(K)$, $\overline{T} \subseteq \mathcal{D}_n(K)$. If $a \in M$, then let $\phi(a) \in \mathcal{D}_n(K)$ denote the diagonal matrix with the same diagonal as a. By Corollary 4.8, $\phi(G) = \phi(T) = T$. So $0 = \phi(0) \in \phi(M) = \phi(\overline{G}) \subseteq \overline{\phi(G)} = \overline{T}$.

Lemma 6.5. Let M be an algebraic monoid with group of units G. Then $M^c g = gM^c$ for all $g \in G$ and $\overline{G} = M^c G = GM^c$.

Proof. Since $G^c \triangleleft G$ and $M^c = \overline{G^c}$, it follows that $g^{-1}M^c g = M^c$ for all $g \in G$. Since G/G^c is a finite group, there exist $g_1,\ldots,g_k \in G$ such that $G = G^c g_1 \cup \ldots \cup G^c g_k$. Then

$$\overline{G} = M^c g_1 \cup \ldots \cup M^c g_k \subseteq M^c G \subseteq \overline{G}G = \overline{G}$$

This proves the lemma.

The next result is due to the author [73; Lemma 1.1].

Lemma 6.6. Let M be a connected monoid with group of units G, I a closed connected right ideal of M, $e \in E(I)$. Let $\dim I = n$, $\dim eM = m$. Then every irreducible component of the closed set $\{a \in I \mid ea = e\}$ has dimension $n - m$.

Proof. Since I is a right ideal of M, $eM \subseteq I$. Let $Y = \{a \in eM \mid a \not\mathcal{J} e \text{ in } M\}$. Then Y is a closed set by Lemma 3.27. So $V = eM \backslash Y$ is a non-empty open subset of eM. Consider the surjective morphism $\phi: I \to eM$ given by $\phi(a) = ea$. By Theorem 2.21, there exists a non-empty open subset U of eM such that every irreducible component of $\phi^{-1}(u)$, $u \in U$ has dimension $n - m$. Since eM is a connected semigroup, $V \cap U \neq \emptyset$. Let $u \in V \cap U$. Then $eu = u$, $e \mathcal{J} u$. By Theorem 1.4 (i), $e \mathcal{R} u$. By Proposition 6.1, $eg = u$ for some $g \in G$. Let $\phi^{-1}(u) = A_1 \cup \ldots \cup A_t$ represent the decomposition of $\phi^{-1}(u)$ into irreducible components. Then $\dim A_i = n - m$, $i = 1,\ldots,t$. If $a \in I$, then $ea = e$ if and only if $eag = u$. Since I is a right ideal of M, $Ig = I$. It follows that $\phi^{-1}(e) = \phi^{-1}(u)g^{-1} = A_1g^{-1} \cup \ldots \cup A_tg^{-1}$. This proves the lemma.

Recall that $M_e = \{a \in M \mid ae = ea = e\}^c$, $G_e = G \cap M_e$. The following result and its corollaries are due to the author [66], [67].

Theorem 6.7. Let M be a connected monoid with $e \in E(M)$. Then $E(M_e) = \{f \in E(M) \mid f \geq e\}$ and e is the zero of M_e.

Proof. We may assume that $e \neq 1$. Let G denote the group of units of M, $S = M \backslash G$, $\dim M = n$. Let $S_1,\ldots,S_k$ denote the irreducible components of S. Then by Proposition 6.2, each S_i is an ideal of M and $\dim S_i = n - 1$, $i = 1,\ldots,k$. Let $M_1 = \{a \in M \mid ea = e\}$. Let X be an irreducible component of M_1 containing e. Then by Proposition 6.6, $\dim X = n - \dim e M$. Suppose $X \subseteq S$. Then $X \subseteq S_i$ for some i. By Proposition 6.6, $\dim X = n - 1 - \dim eM$, a contradiction. Thus $X \cap G \neq \emptyset$. Choose $g \in X \cap G$. Then $eg = e$. So $1, e \in Xg^{-1}$. Hence $Xg^{-1} \subseteq M_1^c$. So $e \in M_1^c$. Let $M_2 = \{a \in M_1^c \mid ae = e\}$. Then $M_2^c = M_e$. By the dual of the above argument, $e \in M_e$. Now let $f \in E(M)$, $f \geq e$. Then $G_f \subseteq G_e$. So $f \in M_f \subseteq M_e$.

Corollary 6.8. Let M be a connected monoid with group of units G, $e, f \in E(M)$. Then

(i) $e \mathscr{J} f$ if and only if $x^{-1}ex = f$ for some $x \in G$.

(ii) $e \mathscr{R} f$ if and only if there exists $x \in G$ such that $ex = x^{-1}ex = f$.

(iii) $e \mathscr{L} f$ if and only if there exists $x \in G$ such that $xe = xex^{-1} = f$.

Proof. By Theorem 5.9, (i) will follow from (ii), (iii). By symmetry it suffices to prove (ii). So let $e \mathscr{R} f$. Let $M_1 = \{a \in M \mid ae = e\}$. Then $G_e, G_f \subseteq M_1^c$. By Theorem 6.7, $e, f \in M_1^c$. Since $e \mathscr{R} f$, we see by Proposition 6.1 that there exists $x \in M_1^c \cap G$ such that $ex = f$. Then $xe = e$ and $x^{-1}ex = f$. The result follows.

Corollary 6.9. Let M be a connected monoid, $J_1, J_2 \in \mathscr{U}(M)$. Then the following conditions are equivalent.

(i) $J_1 \geq J_2$.

(ii) For all $e_1 \in E(J_1)$ there exists $e_2 \in E(J_2)$ such that $e_1 \geq e_2$.

(iii) For all $e_2 \in E(J_2)$ there exists $e_1 \in E(J_1)$ such that $e_1 \geq e_2$.

Proof. That (ii) => (i), (iii) => (i) is obvious. Let G denote the group of units of M and suppose $J_1 \geq J_2$. Let $e \in E(J_1)$, $f \in E(J_2)$. Then for some $x, y \in M$, $xey = f$. Let $f' = eyfxe \in E(J_2)$. Then $e \geq f'$. By Corollary 6.8, there exists $x \in G$ such that $f = x^{-1}f'x$. Let $e' = x^{-1}ex \in E(J_1)$. Then $e' \geq f$.

Corollary 6.10. Let M be a connected monoid with group of units G. Then

(i) For any chain $\Gamma \subseteq E(M)$, there exists a maximal torus T of G such that $\Gamma \subseteq E(\overline{T})$.

(ii) For any maximal torus T of G, $E(M) = \bigcup_{x \in G} x^{-1}E(\overline{T})x$.

Proof. Since all maximal tori are conjugate, (ii) follows from (i). So we prove (i) by induction on $|\Gamma|$. If $|\Gamma| = 0$, there is nothing to prove. So let $|\Gamma| \geq 1$. Let e be the smallest element of Γ. By Theorem 6.7, $\Gamma \subseteq \dot{M}_e$ and e is the zero of M_e. There exists a maximal torus T_1 of G_e such that $\Gamma \backslash \{e\} \subseteq \overline{T}_1$. By Corollary 6.4, $e \in \overline{T}_1$. Thus $\Gamma \subseteq \overline{T}$ for any maximal torus T of G containing T_1.

The following result is due to the author [73; Theorem 1.3].

Theorem 6.11. Let M be a connected monoid with group of units G, $e \in E(M)$. Let $M_1 = \{a \in M | ea = e\}$, $M_2 = \{a \in M | ae = e\}$, $M_3 = \{a \in M | ea = ae = e\}$. Let $G_i = M_i \cap G$, $i = 1,2,3$. Then $M_i = \overline{G}_i$, $i = 1,2,3$.

Proof. First we show that $M_1 = \overline{G}_1$. Let $\dim M = n$, $\dim eM = q$. Let $a \in M_1$, X an irreducible component of M containing a. Then by Lemma 6.6, $\dim X = n - q$. Suppose $X \subseteq S = M \backslash G$. Then $X \subseteq S'$ for some irreducible component S' of S. By Proposition 6.2, S' is an ideal of M, $\dim S' = n - 1$. Now $e = ea \in S'$. So by Lemma 6.6, $\dim X = n - 1 - q$, a contradiction. So $x \cap G \neq \emptyset$. Thus $a \in X = \overline{X \cap G} \subseteq \overline{G}_1$. Thus $M_1 = \overline{G}_1$. Similarly, $M_2 = \overline{G}_2$.

Now let $a \in M_3 \subseteq M_1$. By Lemma 6.5, there exists $g \in G_1$ such that $a \in M_1^c g = gM_1^c$. Now $e \in M_1^c$ by Theorem 6.7. So $e = eg \in M_1^c g = gM_1^c$. Hence $f = g^{-1}e \in M_1^c$. So $fe = f$, $ef = e$. Thus $f \in E(M_1^c)$, $e \mathscr{L} f$. By Proposition 6.1, $f = ue$ for some $u \in G_1^c$. So $u^{-1}g^{-1}e = e = eu^{-1}g^{-1}$. Since $u^{-1}g^{-1} \in G_3$ and $a \in M_3$, we see that $u^{-1}g^{-1}a \in M_1^c \cap M_3 = M_1^c \cap M_2$. Considering the monoid M_1^c, we see that $M_1^c \cap M_2 = \overline{G_1^c \cap G_2} \subseteq \overline{G}_3$. So $u^{-1}g^{-1}a \in \overline{G}_3$. Thus $a \in \overline{G}_3$, proving the theorem.

Example 6.12. Let $M = \{(a,b,c) \mid a,b,c \in K, a^2b = c^2\}$ with pointwise multiplication. Then M is a connected monoid with zero. Let G denote the group of units of M, $e = (0,1,0) \in E(M)$. Then $G_1 = \{x \in G \mid ex = e\}$ is not connected.

The following result is due to the author [73; Theorem 1.4].

Corollary 6.13. Let M be a connected monoid with group of units G and let $e \in E(M)$. Then $GM_eG = \{a \mid a \in M, a \mid e\}$.

Proof. Let $a \in M$, $a \mid e$. Choose $e_1 \in E(M)$, $a \mid e_1 \mid e$ such that $J = J_{e_1}$ is maximal. Now $xay = e_1$ for some $x,y \in M$. Let $z = ye_1x$, $e_2 = az$. Then $e_2 \in E(M)$, $xe_2ay = e_1$. So $e_1 \mathcal{J} e_2$. Now $e_2az = e_2$. Thus $e_2a \mathcal{R} e_2$. By Proposition 6.1, $e_2a = e_2w$ for some $w \in G$. Let $b = aw^{-1}$. Then $a \mathcal{J} b$, $e_2b = e_2$. Now $b^k \mathcal{H} f$ for some $f \in E(M)$, $k \in \mathbb{Z}^+$. Then $e_2 = e_2b^k = e_2b^kf$. So $e_2f = e_2$. Thus $a \mid b \mid f \mid e_2 \mid e_1$. By the maximality of J, $e_2 \mathcal{J} f$. So $e_2 \mathcal{L} f$. Since $e_2b = e_2$, $f = fe_2 = fe_2b = fb$. So $fb^k = f$. Since $b^k \mathcal{H} f$, $b^k = f$. So $fb = bf = f$. Since $f \mid e$, there exists by Corollary 6.9, $e' \in E(M)$ such that $f \geq e'$, $e' \mathcal{J} e$. So $be' = e'b = e'$. By Corollary 6.8, $y^{-1}e'y = e$ for some $y \in G$. So $y^{-1}bye = ey^{-1}by = e$. By Lemma 6.5, Theorem 6.11, $y^{-1}by \in M_eG$. Since $b = aw^{-1}$, $a \in GM_eG$, proving the result.

Lemma 6.14. Let M be a commutative connected monoid, $e \in E(M)$. Let Ω be a finite group of automorphisms of M having e as a common fixed point. Then there exists a closed connected submonoid M_1 of M such that $e \in M_1$ and $a^\sigma = a$ for all $\sigma \in \Omega$, $a \in M_1$.

Proof. Let $\Omega = \{\sigma_1,...,\sigma_p\}$. Define a homomorphism $\psi\colon M \to M$ as $\psi(x) = x^{\sigma_1} ... x^{\sigma_p}$. Then $\psi(1) = 1$, $\psi(e) = e$. Let $\sigma \in \Omega$. Then $\Omega\sigma = \Omega$, whereby $\psi(x)^\sigma = \psi(x)$ for all $x \in M$. Let $M_1 = \overline{\psi(M)}$.

Example 6.15. Let $M = \{A \otimes B \mid A,B \in \mathcal{M}_2(K)\}$, $e = \begin{bmatrix}1 & 0\\0 & 0\end{bmatrix} \otimes \begin{bmatrix}1 & 0\\0 & 0\end{bmatrix}$, $f = \begin{bmatrix}1 & 1\\0 & 0\end{bmatrix} \otimes \begin{bmatrix}0 & 0\\0 & 1\end{bmatrix}$. Then $ef = fe = 0$ but $f \notin \overline{C_G(e)}$. Thus $C_M(e)$ is not a connected monoid.

The following result is due to the author [67], [69].

Theorem 6.16. Let M be a connected monoid with group of units G and let $e \in E(M)$. Then

(i) $C_G^r(e)$, $C_G^\ell(e)$, $C_G(e)$ are closed, connected subgroups of G.

(ii) $eM \subseteq \overline{C_G^r(e)}$, $Me \subseteq \overline{C_G^\ell(e)}$, $eMe \subsetneq \overline{C_G(e)}$.

(iii) If H denotes the $\mathcal{H}$-class of e, then the map $\gamma\colon C_G(e) \to H$ given by $\gamma(a) = ea$ is a surjective homomorphism and $|W(C_G(e))| = |W(H)|.|W(G_e)|$. If T is a maximal torus of G with $e \in \overline{T}$, then T_e, eT are maximal tori of G_e, H, respectively.

Proof. (i) By Corollary 6.10, $e \in E(\overline{T})$ for some maximal torus T of G. First we show that $C_G(e)$ is a connected group. Let $x \in C_G(e)$. Then T, $x^{-1}Tx$ are maximal tori of $C_G(e)^c$. So $y^{-1}x^{-1}Txy = T$ for some $y \in C_G(e)^c$. So $u = xy \in N_G(T) \cap C_G(e)$. Consider the automorphism $\sigma\colon \overline{T} \to \overline{T}$ given by $a^\sigma = u^{-1}au$. Since $W = N_G(T)/C_G(T)$ is a finite group, σ is of finite order. Clearly $e^\theta = e$ for all $\theta \in \langle \sigma \rangle$. So by Lemma 6.14, there exists a closed connected subgroup T_1 of T such that $e \in \overline{T_1}$, $a^\sigma = a$ for all $a \in T_1$. So $u \in C_G(T_1) = C_G(\overline{T_1}) \subseteq C_G(e)$. But $C_G(G_1)$ is connected by Theorem 4.11 (iv). Hence $C_G(T_1) \subseteq C_G(e)^c$. Thus $xy = u \in C_G(e)^c$. Since $y \in C_G(e)^c$, $x \in C_G(e)^c$. So $C_G(e) = C_G(e)^c$ is a connected group.

Now let $x \in C^r_G(e)$. Then T, $x^{-1}Tx$ are maximal tori of $C^r_G(e)^c$. Hence $y^{-1}x^{-1}Txy = T$ for some $y \in C^r_G(e)^c$. So $u = xy \in N_G(T) \cap C^r_G(e)$. Clearly $N_G(T) = N_G(\overline{T})$. So $f = ueu^{-1} \in \overline{T}$. But $ue = eue$. So $ef = f$. Since $\overline{T}$ is commutative, $e = f$. So $u \in C_G(e) \subseteq C^r_G(e)^c$. Hence $x \in C^r_G(e)^c$. Thus $C^r_G(e) = C^r_G(e)^c$ is a connected group. Similarly $C^\ell_G(e)$ is a connected group.

(ii) We may assume that $e \neq 1$. Let $\dim M = n$, $S = M\backslash G$, $S_1,\ldots,S_k$ the irreducible components of S. Then by Proposition 6.2, $\dim S_i = n - 1$, $i = 1,\ldots,k$. Let $\dim eM = q < n$. Define $\phi\colon M \to eM$ as $\phi(a) = ea$. Let ϕ_i denote the restriction of ϕ to S_i, $V_i = \overline{\phi(S_i)}$, $i = 1,\ldots,k$. Let $i \in \{1,\ldots,k\}$. If $V_i \neq eM$, let $U_i = eM\backslash V_i$. Next suppose $V_i = eM$. Then ϕ_i is a dominant morphism. So by Theorem 2.21, there exists a non–empty open set U_i of eM contained in $\phi(S_i)$, such that for any closed irreducible subset Y of eM with $Y \cap U_i \neq \emptyset$, any irreducible component X of $\phi_i^{-1}(Y)$ with $\phi(X) = Y$, we have

$$\dim X = \dim Y + n - 1 - q \tag{3}$$

Let $U = U_1 \cap \ldots \cap U_k$. Since eS is irreducible, U is a non–empty open subset of eS. Let $x \in U$. Then $x \in \phi^{-1}(x)$. Let F be an irreducible component of $\phi^{-1}(x)$ with $x \in F$. By Theorem 2.21, $\dim F \geq n - q$. Suppose $F \subseteq S$. Then $F \subseteq S_i$ for some i. Hence $x = \phi(x) \in \phi(S_i) \subseteq V_i$. Since $x \in U_i$, we see that $U_i \cap V_i \neq \emptyset$. So $V_i = eM$. By (3), $\dim F = n - 1 - q < n - q$, a contradiction. So there exists $g \in G \cap F$. Then $eg = x$. Since $g \in G$, $Y = eMx = eMeg$ is a closed irreducible subset of eM and $Mx = Meg$ is a closed irreducible subset of $\phi^{-1}(Y)$. Let X be an irreducible component of $\phi^{-1}(Y)$ containing Mx. Then $\phi(X) = Y$. So by Theorem 2.21,

$$\dim X \geq \dim Y + n - q \tag{4}$$

Suppose $X \subseteq S$. Then $X \subseteq S_i$ for some i. So X is an irreducible component of $\phi_i^{-1}(Y)$. Also, $x \in Y \cap U \subseteq Y \cap U_j$. Thus we have a contradiction by (3), (4). So $X \cap G \neq \emptyset$. Let $X_1 = Xg^{-1}$. Then $Me \subseteq X_1$, $X_1 \cap G \neq \emptyset$, X_1 is a closed irreducible subset of M. So $X_1 = \overline{X_1 \cap G}$. Let $a \in X_1 \cap G$. Then $ag \in X$. So $eag = \phi(ag) \in eMeg$ and $ea = eae$. Thus $a \in C^{\ell}_G(e)$. So $X_1 \cap G \subseteq C^{\ell}_G(e)$. Hence $Me \subseteq X_1 = \overline{X_1 \cap G} \subseteq \overline{C^{\ell}_G(e)}$. Similarly $eM \subseteq \overline{C^{r}_G(e)}$. Applying this result to the connected monoid $\overline{C^{\ell}_G(e)}$, we see that $eMe \subseteq e\overline{C^{\ell}_G(e)} \subseteq \overline{C_G(e)}$.

(iii) Clearly γ is a homomorphism. By Corollary 4.3 (i), $\gamma(C_G(e))$ is closed in H. In M,

$$H \subseteq eMe = e\overline{C_G(e)} = \gamma(\overline{C_G(e)}) \subseteq \overline{\gamma(C_G(e))}$$

It follows that $H = \gamma(C_G(e))$. Clearly $(\ker \gamma)^c = G_e$. So we are done by Corollaries 4.8, 4.16, 4.25.

Corollary 6.17. Let M be a connected monoid with group of units G and let T be a maximal torus of G. Let $a \in M$, $\Gamma \subseteq E(\overline{T})$. Then

(i) $C^r_G(\Gamma)$, $C^{\ell}_G(\Gamma)$, $C_G(\Gamma)$ are closed connected subgroups of G.

(ii) If $ea = eae \mathcal{H} e$ for all $e \in \Gamma$, then $a \in \overline{C^{\ell}_G(\Gamma)}$. If $ae = eae \mathcal{H} e$ for all $e \in \Gamma$, then $a \in \overline{C^{r}_G(\Gamma)}$. If $ea = ae \mathcal{H} e$ for all $e \in \Gamma$, then $a \in \overline{C_G(\Gamma)}$.

Proof. (i) follows from the repeated application of Theorem 6.16 (i). So we prove (ii). Suppose $ea = eae \mathcal{H} e$ for all $e \in \Gamma$. We prove by induction on $|\Gamma|$ that $a \in \overline{C^{\ell}_G(\Gamma)}$. If $|\Gamma| = 0$, this is clear. So let $e \in \Gamma$, $\Gamma' = \Gamma \backslash \{e\}$. Let $G' = C^{\ell}_G(\Gamma')$,

$M' = \overline{G'}$. Then $T \subseteq G'$, $a \in M'$. Now $ea = eae \mathscr{H} e$ in M and hence in M' by Remark 1.3 (iii). By Theorem 6.16 (iii), there exists $u \in C_{G'}(e)$ such that $ea = eae = eu$. Then $eau^{-1} = e$. Let $M_1 = \{b \in M' \mid eb = e\}$, $G_1 = M_1 \cap G'$. Then by Theorem 6.11, $M_1 = \overline{G_1}$. Clearly $G_1 \subseteq C^{\ell}_{G'}(e)$. Hence $au^{-1} \in \overline{C^{\ell}_{G'}(e)}$. Thus $a \in \overline{C^{\ell}_{G'}(e)} = \overline{C^{\ell}_{G}(\Gamma)}$. The other statements are proved similarly.

<u>Corollary 6.18</u>. Let M be a connected monoid with group of units G and let Γ be a chain in $E(M)$. Then

(i) $C^{r}_{G}(\Gamma)$, $C^{\ell}_{G}(\Gamma)$, $C_{G}(\Gamma)$ are connected groups.

(ii) If $e \in \Gamma$, $f \in E(M)$, then $e \mathscr{R} f$ implies $f \in \overline{C^{r}_{G}(\Gamma)}$.

(iii) If $e \in \Gamma$, $f \in E(M)$, then $e \mathscr{L} f$ implies $f \in \overline{C^{\ell}_{G}(\Gamma)}$.

<u>Proof</u>. By Corollary 6.10, $\Gamma \subseteq E(\overline{T})$ for some maximal torus T of G. Thus (i) follows from Corollary 6.17. Since (iii) is dual to (ii), it suffices to prove (ii). We proceed by induction on $|\Gamma|$. If $|\Gamma| = 0$, there is nothing to prove. So let $|\Gamma| > 0$. If $h \in \Gamma$, let $X_h = \{f \in E(M) \mid f \mathscr{R} h\}$. Let e be the maximum element of Γ, $\Gamma' = \Gamma \backslash \{e\}$. Let $G_1 = C^{r}_{G}(e)$, $M_1 = \overline{G_1}$. By Theorem 6.16 (ii), $e M \subseteq M_1$. Thus $X_h \subseteq e M \subseteq M_1$ for all $h \in \Gamma$. Let $G_2 = C^{r}_{G}(\Gamma) = C^{r}_{G_1}(\Gamma')$, $M_2 = \overline{G_2}$. By the induction hypothesis, $X_h \subseteq M_2$ for all $h \in \Gamma'$. Now let $f \in X_e$. Then $e \mathscr{R} f$. Let $a \in G_f$. Then $af = f$. So for all $h \in \Gamma$, $ah = afeh = h$. Hence $a \in C^{r}_{G}(\Gamma)$. So $G_f \subseteq C^{r}_{G}(\Gamma)$. By Theorem 6.7, $f \in \overline{G_f} \subseteq \overline{C^{r}_{G}(\Gamma)}$. This completes the proof.

<u>Corollary 6.19</u>. Let M be a connected monoid, $e,f \in E(M)$, $e \mathscr{J} f$. Then $C^{r}_{G}(e) = C^{r}_{G}(f)$ if and only if $e \mathscr{R} f$; $C^{\ell}_{G}(e) = C^{\ell}_{G}(f)$ if and only if $e \mathscr{L} f$.

Proof. First suppose $e \mathcal{R} f$. By Corollary 6.18 (ii), $e,f \in \overline{C_G^r(e)}$. By Corollary 6.8, $x^{-1}ex = f$ for some $x \in C_G^r(e)$. So $C_G^r(e) = x^{-1}C_G^r(e)x = C_G^r(x^{-1}ex) = C_G^r(f)$. Next, suppose $C_G^r(e) = C_G^r(f)$. Since $e \mathcal{J} f$, we see by Theorem 5.9 that there exists $e', f' \in E(M)$ such that $e \mathcal{R} e' \mathcal{L} f' \mathcal{R} f$. By the above, $C_G^r(e') = C_G^r(f')$. By Corollary 6.18, $e',f' \in \overline{C_G^r(e')}$. By Corollary 6.8, there exists $y \in C_G^r(e')$ such that $ye'y^{-1} = f'$. But $ye' = e'ye'$. So $f' = e'f' = e'$. Hence $e \mathcal{R} f$.

The next result is from the author [65], [66].

Theorem 6.20. Let M be a connected monoid with group of units G. Let T be a maximal torus of G. Then $E(\overline{T})$ is a finite, relatively complemented lattice. Moreover, the lengths of the maximal chains in $E(\overline{T})$, $E(M)$, $\mathcal{U}(M)$ are all the same. If M has a zero, then this number is equal to $\dim T$.

Proof. $E(\overline{T}) \cong \mathcal{U}(\overline{T})$ is a finite lattice by Theorem 5.10. Let J_o denote the kernel of M. Then by Corollary 6.10, $J_o \cap E(\overline{T}) = \{\nu\}$ where ν is the zero of $E(\overline{T})$. By Theorem 6.7, $E(\overline{T}) = E(\overline{T}_\nu)$. Also, by Corollary 6.9, any maximal chain in $\mathcal{U}(M)$ gives rise to a maximal chain in $E(M_\nu)$. Thus we may assume that $\nu = 0$ is the zero of M. Let $\Gamma = \{1 > e > \ldots > 0\}$ be a maximal chain in $E(\overline{T})$. We prove by induction on $|\Gamma|$ that $\dim T = |\Gamma| - 1$. Now eT is the group of units of $\overline{eT} = e\overline{T}$ and $\Gamma' = \Gamma \backslash \{1\}$ is a maximal chain in $e\overline{T}$. Thus $\dim eT = |\Gamma'| - 1 = |\Gamma| - 2$. We have a surjective homomorphism $\phi: T \to eT$ given by $\phi(t) = et$. Clearly $(\ker \phi)^c = T_e$, $\overline{T}_e = T_e \cup \{e\}$. By Proposition 6.2, $\dim T_e = 1$. By Corollary 4.3 (ii), $\dim T = \dim eT + 1 = |\Gamma| - 1$.

By Corollary 6.9, a maximal chain in $\mathcal{U}(M)$ gives rise to a maximal chain in $E(M)$. By Corollary 6.10, a maximal chain in $E(M)$ is contained in the closure of some maximal torus and hence by the above, has length equal to $\dim T$.

Finally, we show that $E(\overline{T})$ is relatively complemented. Let $e_1,e,f \in E(\overline{T})$, $e_1 > e > f$. We need to find $h \in E(\overline{T})$ such that $e_1 > h > f$, $eh = f$, $e \vee h = e_1$. We may assume that $e_1 = 1$ (otherwise we work with $e_1\overline{T}$). Consider the homomorphism $\phi: \overline{T} \to e\overline{T}$ given by $\phi(a) = ea$. Now $\dim e\overline{T} < \dim \overline{T}$, $\phi(f) = f$. By Theorem 2.21, $\dim \phi^{-1}(f) > 0$. So there exists $x \in \overline{T}$, $x \neq f$ such that $ex = f$. Now $x \mathcal{H} h$ for some $h \in E(\overline{T})$. Then $eh = f$. If $h = f$, then $x = fx = efx = ex = f$, a contradiction. Choose $h \in E(\overline{T})$ maximal with $eh = f$. We claim that $e \vee h = 1$. For suppose $h_1 = e \vee h \neq 1$. Then by the above, there exists $h_2 \in E(\overline{T})$ such that $h_2 > h$, $h_1h_2 = h$. Then $eh_2 = eh_1h_2 = eh = f$, contradicting the maximality of h. Thus $e \vee h = 1$, proving the theorem.

<u>Definition 6.21</u>. Let M be a connected monoid with kernel J_o. Theorem 6.20 gives rise to <u>height function</u>, ht on $\mathcal{U}(M)$, $E(M)$ as follows: $ht(J_o) = 0$, if $J, J' \in \mathcal{U}(M)$ with J covering J', then let $ht(J) = ht(J') + 1$. If $J \in \mathcal{U}(M)$, $e \in E(J)$, let $ht(e) = ht(J)$. If $ht(1) = p$, let $ht(M) = ht(E(M)) = p$. For $i = 0,\ldots,p$, let

$$\mathcal{U}_i(M) = \mathcal{U}^{(p-i)}(M) = \{J \in \mathcal{U}(M) \mid ht(J) = i\}.$$

<u>Corollary 6.22</u>. Let M be a connected monoid with group of units G and let T be a maximal torus of G. Let $e_1,e_2 \in E(\overline{T})$, $e_1 > e_2$, J_i the $\mathcal{J}$-class of e_i, $i = 1,2$. Then the following conditions are equivalent.

(i) e_1 covers e_2 in $E(\overline{T})$.

(ii) e_1 covers e_2 in $E(M)$.

(iii) J_1 covers J_2 in $\mathcal{U}(M)$.

<u>Proof</u>. Clearly (iii) => (ii) => (i). So assume (i). Now $e_1,e_2 \in \Gamma$ for some maximal chain Γ of $E(\overline{T})$. By Theorem 6.20, $J_e(e \in \Gamma)$ is a maximal chain in $\mathcal{U}(M)$. Hence J_1 covers J_2.

Corollary 6.23. Let S be a connected semigroup. Then the length of any maximal chain in $\mathcal{U}(S)$ = the length of any maximal chain in $E(S)$.

Proof. By Theorem 5.10, $\mathcal{U}(S)$ has a maximum element J_o. Fix $e \in E(J_o)$. Let Ω be a maximal chain in $\mathcal{U}(S)$. Then $\Omega' = \{J \cap eSe \mid J \in \Omega\}$ is a maximal chain in $\mathcal{U}(eSe)$. We can now apply Theorem 6.20.

The following result is due to the author [68].

Proposition 6.24. Let M be a connected monoid with group of units G such that a maximal subgroup of the kernel of M is solvable. Then for any maximal chain Γ of $E(M)$, $C_G^r(\Gamma)$, $C_G^\ell(\Gamma)$ are connected solvable groups.

Proof. By Corollary 6.18, $C_G^r(\Gamma)$, $C_G^\ell(\Gamma)$ are connected. We prove by induction on $|\Gamma|$ that $C_G^r(\Gamma)$ is solvable. If $|\Gamma| = 1$, there is nothing to prove. So let $|\Gamma| > 1$. Let e be the maximum element of $\Gamma' = \Gamma\backslash\{1\}$. Then 1 covers e. Let H denote the $\mathcal{H}$-class of e. Clearly Γ' is a maximal chain in $E(eMe)$. Hence $C_H^r(\Gamma')$ is a solvable group. Define ϕ: $C_G^r(\Gamma) \to H$ as $\phi(a) = ae = eae$. Clearly ϕ is a homomorphism, $\phi(C_G^r(\Gamma)) \subseteq C_H^r(\Gamma')$ is solvable. Clearly, the kernel of ϕ, $G_1 = \{a \in G \mid ae = e\}$. Let $M_1 = \overline{G_1^c}$ and let T be a maximal torus of G_1^c. Then $E(\overline{T}) = \{1,e\}$, e is the zero of $\overline{T}$. So by Theorem 6.20, $\dim T = 1$. We may assume that M_1 is a closed submonoid of some $\mathcal{M}_n(K)$. Then $\det a = 1$, for all $a \in G_2 = (G_1^c, G_1^c)$. So G_2 is closed in M_1. Since $e \in \overline{T}$, $T \not\subseteq G_2$. Hence G_2 is a unipotent group and therefore solvable. It follows that G_1^c is solvable. By Theorem 4.22 (iv), Corollary 4.25, $C_G^r(\Gamma)$ is a solvable group.

Proposition 6.25. Let M be a connected monoid with group of units G and let T be a maximal torus of G. Let $J, J' \in \mathcal{U}(M)$, $J \geq J'$. Let $A = J \cap E(\overline{T})$, $A' = J' \cap E(\overline{T})$.

Then

(i) For any $e \in A$, $A = \{e^\sigma \mid \sigma \in W\}$ and $|W| = |A| \cdot |W(C_G(e))|$.

(ii) For any $e \in A$, there exists $e' \in A'$ such that $e \geq e'$.

(iii) For any $e' \in A'$, there exists $e \in A$ such that $e \geq e'$.

Proof. (i) Let $e, f \in A$ such that $e \mathcal{J} f$. By Corollary 6.8, $x^{-1}ex = f$ for some $x \in G$. Since $T \subseteq C_G(f)$, $xTx^{-1} \subseteq C_G(e)$. Thus T, xTx^{-1} are maximal tori of $C_G(e)$. So there exists $y \in C_G(e)$ such that $T = yxTx^{-1}y^{-1}$. Then $yx \in N_G(T)$. Let $\sigma = yxC_G(T)$. Then $e^\sigma = x^{-1}y^{-1}eyx = x^{-1}ex = f$. Thus W acts transitively on A. Clearly $e^\sigma = e$ if and only if $\sigma \in W(C_G(e))$.

(ii) Let $e \in A$, H the $\mathcal{H}$-class of e. By Theorem 6.16, eT is a maximal torus of H. By Corollary 6.9, there exists $f \in E(J')$ such that $e \geq f'$. So $f \in E(eMe)$. By Corollary 6.10, there exists $e' \in E(e\overline{T})$ such that $f \mathcal{J} e'$. Then $e \geq e'$, $e' \in A'$.

(iii) Let $e' \in A'$. By Corollary 6.9 there exists $e_1 \in J$ such that $e_1 \geq e'$. By Theorem 6.16, $T_{e'}$ is a maximal torus of $G_{e'}$. Since $e_1 \in M_{e'}$, we see by Corollary 6.10 that there exists $e \in E(\overline{T}_{e'})$ such that $e_1 \mathcal{J} e$. Then $e \geq e'$, $e \in A$.

Let M be a connected monoid with group of units G and let T be a maximal torus of G.

Definition 6.26. If $J \in \mathcal{U}(M)$, then the width of J, $w(J) = |J \cap E(\overline{T})|$. If $e \in E(J)$, the width of e, $w(e) = w(J)$.

Since all maximal tori in G are conjugate, the above definition is independent of T. Note that the width of e in $\overline{C_G^r(e)}$ and $\overline{C_G^\ell(e)}$ is 1 by Corollary 6.8. Also note that if $e \in E(\overline{T})$, then by Proposition 6.25 (i),

$$w(e) = |\{e^{\sigma} | \sigma \in W\}| .$$

The following result is due to the author [70].

Proposition 6.27. Let M be a connected monoid with group of units G and let $e,f,f' \in E(M)$, $f \mathcal{J} f'$. Let J_0 denote the kernel of M, J the $\mathcal{J}$-class of e. Then

(i) If $e \geq f, f'$, then $f,f' \in eMe$ and $f \mathcal{J} f'$ in eMe.

(ii) If $e \leq f, f'$, then $f,f' \in M_e$ and $f \mathcal{J} f'$ in M_e.

(iii) $\mathcal{U}(eMe) \cong [J_0,J]$ and $\mathcal{U}(M_e) \cong [J,G]$.

Proof. (i) is obvious and (iii) follows from (i), (ii). So we proceed to prove (ii). Thus assume that $e \leq f,f'$. Then by Theorem 6.7, $f,f' \in M_e$. By Corollary 6.8, $x^{-1}f'x = f$ for some $x \in G$. Then $f \geq e$, $f \geq x^{-1}ex$. So $e \mathcal{J} x^{-1}ex$ in $fMf \subseteq \overline{C_G(f)}$. Thus there exists $y \in C_G(f)$ such that $y^{-1}x^{-1}exy = e$. So $xy \in C_G(e)$ and $y^{-1}x^{-1}f'xy = y^{-1}fy = f$. Hence $f \mathcal{J} f'$ in $\overline{C_G(e)}$. Thus we may assume that e lies in the center of M. Let T be a maximal torus of G such that $f \in \overline{T}$. Then $e \in \overline{T}$ and T_e is a maximal torus of G_e. By Corollary 6.10, there exists $f_1 \in E(\overline{T}_e)$ such that $f' \mathcal{J} f_1$ in M_e. Then $f \mathcal{J} f_1$ in M. It suffices to show that $f \mathcal{J} f_1$ in M_e. Let $A = \{h \in E(\overline{T}) | h \mathcal{J} f \text{ in } M\}$, $A_1 = \{h \in E(\overline{T}_e) | h \mathcal{J} f \text{ in } M_e\}$. Then $A_1 \subseteq A$, $f_1 \in A$. So it suffices to show that $|A| = |A_1|$. Let $G_1 = C_{G_e}(f) = C_G(f) \cap G_e$. Then by Theorem 6.16,

$$|W(G)| = |W(G_e)| \cdot |W(eMe)|$$
$$|W(C_G(f))| = |W(G_1)| \cdot |W(eMe)|.$$

By Proposition 6.27, $|W(G)| = |A| \cdot |W(C_G(f)|$, $|W(G_e)| = |A_1| \cdot |W(G_1)|$. It follows that $|A| = |A_1|$.

Lemma 6.28. Let M be a connected monoid with group of units G, $e \in E(M)$, $w(e) = 1$. Then for any Borel subgroups B_1, B_2 of G with $e \in \overline{B}_1 \cap \overline{B}_2$, there exists $y \in C_G(e)$ such that $y^{-1}B_1y = B_2$.

Proof. There exists $x \in G$ such that $x^{-1}B_1x = B_2$. Now $e \in \overline{T}$ for some maximal torus T of B_2. Also $x^{-1}ex \in E(\overline{B}_2)$. By Corollary 6.10, $y^{-1}x^{-1}exy \in \overline{T}$ for some $y \in B_2$. Since $w(e) = 1$, $y^{-1}x^{-1}exy = e$. Thus $xy \in C_G(e)$. Clearly $y^{-1}x^{-1}B_1xy = y^{-1}B_2y = B_2$.

By Lemma 6.14, we have

Lemma 6.29. Let M be a connected monoid with gorup of units G and let T be a maximal torus of G. Let $T_1 = \{t \in T \mid t^\sigma = t$ for all $\sigma \in W\}^c$. Then $E(\overline{T}_1) = \{e \in E(\overline{T}) \mid w(e) = 1\}$.

The following result is due to the author [69], [72].

Theorem 6.30. Let M be a connected monoid with group of units $G, J \in \mathcal{U}(M)$. Then the following conditions are equivalent.

(i) $J^2 = J$ (i.e., $E(J)^2 \subseteq J$).

(ii) $w(J) = 1$.

(iii) $E(J) \subseteq \overline{B}$ for some Borel subgroup B of G.

(iv) $E(J) \subseteq \overline{\text{rad } G}$.

Proof. (i) => (ii). Let T be a maximal torus of G, $e,f \in J \cap E(\overline{T})$. Then $ef = fe \in J^2 \subseteq J$. By Theorem 1.4 (i), $e = f$. Thus $w(J) = 1$.

(ii) => (iii). For $e \in E(J)$, let $X_e = \{f \in E(M) \mid e\ \mathcal{R}\ f\}$, $Y_e = \{f \in E(M) \mid e\ \mathcal{L}\ f\}$. Let $e \in E(J)$, $G_1 = \{a \in G \mid ae = e\}^c$, $M_1 = \overline{G_1}$. By Theorem 6.7, $X_e \subseteq M_1$. Clearly X_e is the kernel of M_1. Let Γ be a maximal chain in $E(M_1)$ with $e \in \Gamma$. Then $G_2 = C_G^r(\Gamma)$ is solvable by Proposition 6.24. By Corollary

6.18, $X_e \subseteq \overline{G_2}$. Now $G_2 \subseteq B$ for some Borel subgroup B of G. Thus $X_e \subseteq \overline{B}$. Similarly there exists a Borel subgroup B_1 of G such that $Y_e \subseteq \overline{B_1}$. Then $e \in \overline{B} \cap \overline{B}_1$, $w(e) = 1$. By Lemma 6.28, there exists $u \in C_G(e)$ such that $u^{-1}B_1u = B$. So $Y_e = u^{-1}Y_eu \subseteq \overline{B}$. Let J' denote the $\mathcal{J}$-class of e in $\overline{B}$. Then for any $h \in E(J')$, there exists $b \in B$ such that $h = b^{-1}eb$. Thus $X_h, Y_h \subseteq \overline{B}$. Now let $f \in E(J)$. Then by Theorem 5.9, there exists $e', f' \in E(M)$ such that $e \mathcal{R} e' \mathcal{L} f' \mathcal{R} f$. Then $e' \in X_e \subseteq \overline{B}$. So $f' \in Y_{e'} \subseteq \overline{B}$. Thus $f \in X_{f'} \subseteq \overline{B}$. So $E(J) \subseteq \overline{B}$.

(iii) => (i). By Theorem 5.19, $E(J)$ is contained in a $\mathcal{J}$-class of $\overline{B}$. By Corollaries 3.20, 4.12, $E(J)^2 \subseteq J$.

Thus the conditions (i), (ii), (iii) are equivalent. Clearly (iv) => (iii). So assume that (i), (ii), (iii) hold. Let $e \in E(J)$. We need to show that $e \in \overline{\text{rad}\, G}$. Now $x^{-1}E(J)x = E(J)$ for all $x \in G$. So $E(J)$ is contained in the closure of every Borel subgroup of G. Now $\text{rad}\, C_G(e)$, being solvable, is contained in some Borel subgroup B_1 of G. If B_2 is any other Borel subgroup of G, then by Lemma 6.28, there exists $y \in C_G(e)$ such that $y^{-1}B_1y = B_2$. So $\text{rad}\, C_G(e) \subseteq B_2$. Thus $\text{rad}\, C_G(e) \subseteq \text{rad}\, G$. Since $G_e \triangleleft C_G(e)$, $\text{rad}\, G_e \subseteq \text{rad}\, G$. Thus it suffices to show that $e \in \overline{\text{rad}\, G_e}$. So we may assume that $e = 0$ is the zero of M. We proceed by induction on $\dim M$. Let T be a maximal torus of G, $T_1 = \{t \in T \mid t^\sigma = t \text{ for all } \sigma \in W\}^c$, $\Gamma = E(\overline{T}_1)$. Then $0 \in \Gamma$ by Lemma 6.29. By Theorem 6.20, Γ is a relatively complemented lattice. Suppose there exists $f \in \Gamma$ such that $f \neq 0,1$. Then there exists $h \in \Gamma$ such that $h \neq 0, 1$, $fh = 0$. Then $f \in \overline{\text{rad}\, G_f} \subseteq \overline{\text{rad}\, G}$. Similarly $h \in \overline{\text{rad}\, G_h} \subseteq \overline{\text{rad}\, G}$. Hence $0 = fh \in \overline{\text{rad}\, G}$ and we are done. So assume $\Gamma = \{0,1\}$. Then $w(f) > 1$ for all $f \in E(M)$ with $f \neq 0,1$. By Remark 4.31, $\text{rad}\, G$ is not unipotent. So $T_2 = T \cap \text{rad}\, G \neq \{1\}$. Let $x \in N_G(T)$, $t \in T_2$. Then $x^{-1}txt^{-1} \in T_2 \cap (G, \text{rad}\, G) \subseteq T_2 \cap \text{rad}_u G = \{1\}$ by Remark 4.33. Thus $t^\sigma = t$ for all $t \in T_2$, $\sigma \in W$. Therefore $T_2 \subseteq T_1$. By Theorem 6.20, $\dim T_1 = 1$. Hence $T_1 = T_2 \subseteq$

$\overline{\text{rad } G}$ and $0 \in \overline{\text{rad } G}$.

Corollary 6.31. Let M be a connected monoid with group of units G. Then $E(\overline{\text{rad } G}) = \{e \in E(M) \mid w(e) = 1\}$. If G is reductive, then $E(\overline{\text{rad } G})$ is a relatively complemented sublattice of $E(\overline{T})$ for any maximal torus T of G.

Proof. The first assertion follows from Theorem 6.30. Now suppose G is reductive. Then $T_o = \text{rad } G$ is a torus lying in the center of G. So for any maximal torus T of G, $T_o \subseteq T$ and $E(\overline{T}_o) = \{e \in E(\overline{T}) \mid e^{\sigma} = e \text{ for all } \sigma \in W\}$ is a relatively complemented sublattice of $E(\overline{T})$ by Theorem 6.20, Lemma 6.29.

Corollary 6.32. Let M be a connected monooid with zero and group of units G. Then the following conditions are equivalent.

(i) G is solvable.

(ii) There exists a maximal chain $\mathcal{V}$ of $\mathcal{U}(M)$ such that $w(J) = 1$ for all $J \in \mathcal{V}$.

(iii) $w(J) = 1$ for all $J \in \mathcal{U}(M)$.

(iv) M is a semilattice of archimedean semigroups.

Proof. (i) => (iv). This follows from Theorem 3.15, Corollaries 3.20, 4.12.

(iv) => (iii). This follows from Corollary 1.16 and Theorem 6.30.

(iii) => (ii). This is obvious.

(ii) => (i). Let T be a maximal torus of G. By Proposition 6.25, we can find a maximal chain Γ of $E(\overline{T})$ such that $\mathcal{V} = \{J_e \mid e \in \Gamma\}$. So $w(e) = 1$ for all $e \in \Gamma$. By Theorem 6.30, $\Gamma \subseteq \overline{\text{rad } G}$. By Corollary 6.10, there exists a maximal torus T_1 of rad G such that $\Gamma \subseteq E(\overline{T}_1)$. By Theorem 6.20, $\dim T_1 = |\Gamma| - 1 = \dim T$. Thus T_1 is a maximal torus of G. Hence G/rad G is a unipotent group. Thus G is solvable.

The following result is due to the author [75].

Theorem 6.33. Let M be a closed connected submonoid of $\mathcal{M}_n(K)$ and let G denote the group of units of M. Let Γ be a maximal chain in $E(M)$. Then $C_G^\ell(\Gamma)C_G^r(\Gamma) = \{a \in G \mid \det_\Gamma(a) \neq 0\} = \{a \in G \mid eae \mathcal{H} e \text{ for all } e \in \Gamma\}$.

Proof. Let $X = \{a \in G \mid \det_\Gamma(a) \neq 0\}$. Let $e \in \Gamma$, H the $\mathcal{H}$-class of e, $c_1 \in C_G^\ell(\Gamma)$, $c_2 \in C_G^r(\Gamma)$. Then $ec_1 = ec_1e \in H$, $c_2e = ec_2e \in H$. So $ec_1c_2e \in H$. Thus $c_1c_2 \in X$ and $C_G^\ell(\Gamma)C_G^r(\Gamma) \subseteq X$. We need to prove that $X \subseteq C_G^\ell(\Gamma)C_G^r(\Gamma)$. Without loss of generality we may assume that $1 \in \Gamma$. We proceed by induction on $|\Gamma|$. If $|\Gamma| = 1$, there is nothing to prove. So let $|\Gamma| > 1$. Let e denote the maximum element of $\Gamma\backslash\{1\}$. Let $\Gamma' = \Gamma\backslash\{e\}$ and let H denote the $\mathcal{H}$-class of e. Let $a \in X$. Then $eae \in H$. By Theorem 6.16 (iii), there exists $x \in C_G(e)$ such that $eae = ex$. Let $G' = C_G(e)$, $M' = \overline{G'}$. By Theorem 6.16 (ii), $\Gamma \subseteq M'$. Let $f \in \Gamma'$, $f \neq 1$. Then $e > f$. So $fxf = fexef = feaef = faf \mathcal{H} f$. By the induction hypothesis there exists $u \in C_{G'}^\ell(\Gamma')$, $v \in C_{G'}^r(\Gamma')$ such that $x = uv$. Then $u \in C_G^\ell(\Gamma)$, $v \in C_G^r(\Gamma)$. Now $eae = ex = euv$. Let $b = u^{-1}av^{-1}$. Since $u,v \in C_G(e)$, we see that $ebe = e$. Let $e_1 = eb \in E(M)$. Then $e \mathcal{R} e_1$. By Theorem 6.16, $e_1 \in \overline{C_G^r(\Gamma)}$. By Proposition 6.1, there exists $y \in C_G^r(\Gamma)$ such that $eb = e_1 = ey$. So $eby^{-1} = e$. Let $f \in \Gamma$, $f \neq 1$. Then $e \geq f$. So $fby^{-1} = f$ and $by^{-1} \in C_G^\ell(\Gamma)$. Thus $b \in C_G^\ell(\Gamma)C_G^r(\Gamma)$. It follows that $a = ubv \in C_G^\ell(\Gamma)C_G^r(\Gamma)$, proving the theorem.

Corollary 6.34. Let M be a connected monoid with group of units G, $e \in E(M)$. Then $w(e) = 1$ if and only if $G = C_G^\ell(e)C_G^r(e)$ if and only if eGe is the $\mathcal{H}$-class of e.

Proof. Let J denote the $\mathcal{J}$-class of e. Suppose w(e) = 1. Then by Theorem 6.30, $eGe \subseteq J$. By Theorem 6.33, $G = C^{\ell}_G(e)C^r_G(e)$. Conversely assume $G = C^{\ell}_G(e)C^r_G(e)$. Then $eGe \subseteq J$. So $J^2 = GeGeG \subseteq GJG = J$. Also by Theorem 6.16, eGe contains the $\mathcal{H}$-class of e. This completes the proof.

By Theorem 6.33, Corollaries 6.32, 6.34, we have,

Corollary 6.35. Let M be a connected monoid with zero and group of units G. Let Γ be a maximal chain in E(M). Then G is solvable if and only if $G = C^{\ell}_G(\Gamma)C^r_G(\Gamma)$.

7 REDUCTIVE GROUPS AND REGULAR SEMIGROUPS

In this chapter we wish to begin to consider the situation when the group of units is reductive. The following result is due to the author [72].

Theorem 7.1. Let M be a connected monoid with zero and reductive group of units G. Let Γ be a maximal chain in $E(M)$. Then $C_G(\Gamma)$ is a maximal torus of G and $C_G^r(\Gamma)$, $C_G^\ell(\Gamma)$ are opposite Borel subgroups of G relative to $C_G(\Gamma)$.

Proof. By Corollary 6.10, $\Gamma \subseteq \overline{T}$ for some maximal torus T of G. Clearly $T \subseteq C_G^r(\Gamma)$. By Proposition 6.24, $C_G^r(\Gamma)$ is a connected solvable group. So $C_G^r(\Gamma) \subseteq B$ for some Borel subgroup B of G. Let B^- denote the opposite Borel subgroup of G relative to T. Then $B \cap B^- = T$. So $\Gamma \subseteq \overline{B^-}$. Now

$$C_{B^-}^r(\Gamma) \subseteq B^- \cap C_G^r(\Gamma) \subseteq B^- \cap B = T \tag{5}$$

By Corollary 6.35 and (5),

$$B^- = C_{B^-}^\ell(\Gamma) C_{B^-}^r(\Gamma) \subseteq C_G^\ell(\Gamma) T = C_G^\ell(\Gamma).$$

Thus $B^- \subseteq C_G^\ell(\Gamma)$. But $C_G^\ell(\Gamma)$ is a connected solvable group by Proposition 6.24. So $B^- = C_G^\ell(\Gamma)$. Thus $C_G^\ell(\Gamma_1)$ is a Borel subgroup for any maximal chain Γ_1 of $E(M)$. Similarly $C_G^r(\Gamma_1)$ is a Borel subgroup for any maximal chain Γ_1 of $E(M)$.

In particular $B = C_G^r(\Gamma)$. Hence $T = B \cap B^- = C_G(\Gamma)$.

Corollary 7.2. Let M be a connected monoid with zero and a reductive group of units G. Let $e \in E(M)$, H the $\mathcal{H}$-class of e. Then

(i) $C_G(e), G_e, H$ are reductive groups.

(ii) If B is a Borel subgroup of G with $e \in E(\overline{B})$, then $C_B(e)$, B_e, $eBe = eC_B(e)$ are Borel subgroups of $C_G(e)$, G_e, H, respectively.

Proof. (i) H is a homomorphic image of $C_G(e)$ by Theorem 6.16. Also $G_e \triangleleft C_G(e)$. So we need only prove that $C_G(e)$ is a reductive group. Now $e \in \Gamma$ for some maximal chain Γ of $E(M)$. By Theorem 7.1, $B_1 = C_G^r(\Gamma)$ is a Borel subgroup of G and hence of $C_G^r(e)$. Clearly the width of e in $\overline{C_G^r(e)}$ is 1. Now $\text{rad}\, C_G(e) \subseteq B_2$ for some Borel subgroup B_2 of $C_G^r(e)$. Then $e \in \overline{B}_1 \cap \overline{B}_2$. By Lemma 6.28, $x^{-1}B_2x = B_1$ for some $x \in C_G(e)$. Hence $\text{rad}\, C_G(e) \subseteq B_1 = C_G^r(\Gamma)$. Similarly $\text{rad}\, C_G(e) \subseteq C_G(\Gamma)$. Thus $\text{rad}\, C_G(e) \subseteq C_G(\Gamma)$, which is a torus by Theorem 7.1.

(ii) Let $T_0 = \text{rad}\, C_G(e)$ which is a torus by (i). By Corollary 6.31, $e \in \overline{T}_0$. Hence $C_G(e) = C_G(T_0)$. Let B be a Borel subgroup of G with $e \in \overline{B}$. Let T be a maximal torus of B with $e \in \overline{T}$. Then $T \subseteq C_G(e)$ and therefore $T_0 \subseteq T$. By Theorem 4.11 (iv), $C_B(e) = C_B(T_0)$ is a Borel subgroup of $C_G(e)$. So by Theorem 6.16, Corollaries 4.8, 4.16, B_e, $eC_B(e)$ are Borel subgroups of G_e, H respectively. By Theorem 6.16, Corollary 6.34, $eC_B(e) = eBe$. This completes the proof.

The following result is due to the author [73] when char $K = 0$ and Renner [94] for arbitrary characteristic. The proof given here is taken from [77].

Theorem 7.3. Let M be a connected monoid with zero 0 and group of units G. Then the following conditions are equivalent.

(i) G is reductive.

(ii) M is regular.

(iii) M has no non–zero nilpotent ideals.

Proof. That (ii) $\Rightarrow$ (iii) is obvious. Assume (iii). By Corollary 6.31, $0 \in \overline{\text{rad } G}$. Suppose $\text{rad } G$ is not a torus. Then by Theorem 5.12, $\overline{\text{rad } G}$ has a non–zero nilpotent element a. Let $x \in M$. Then by Proposition 6.3, $x \in \overline{B}$ for some Borel subgroup B of G. Now $a \in \overline{\text{rad } G} \subseteq \overline{B}$. By Remark 3.15 and Corollary 4.12, we may assume that $\overline{B} \subseteq \mathcal{T}_n(K)$ for some $n \in \mathbb{Z}^+$ and that 0 is the zero matrix. So a is strictly upper triangular. Hence ax is also nilpotent. Thus MaM is a nil ideal of M. It is well–known [35; Chapter VIII, Section 5, Theorem 1] that a nil matrix semigroup is nilpotent. Hence MaM is a non–zero nilpotent ideal of M. This contradiction shows that G is reductive.

Now assume that G is reductive. We prove by induction on $\dim G$ that M is regular. Let $e \in E(M)$, $e \neq 0,1$. Then by Corollary 7.2 and the induction hypothesis eMe and M_e are regular. Let M be a closed submonoid of $\mathcal{M}_n(K)$, $\dim M = p$, $S = M\backslash G$. Let T be a maximal torus of G, B, B^- opposite Borel subgroups of G relative to T, $W = N_G(T)/T$. By Corollary 4.50, we may assume that B, B^- consist of upper and lower triangular matrices, respectively. Then every element of $\overline{B} \cap \overline{B^-}$ is a diagonal matrix. Let S_o be an irreducible component of S. Then by Proposition 6.2, S_o is an ideal of M and $\dim S_o = p - 1$. We first show that S_o is not nil. For suppose otherwise. By Theorem 4.22, $B^- = \sigma^{-1}B\sigma$ for some $\sigma \in W$. By Theorem 6.33, B^-B and hence $B\sigma B$ is an open subset of G. So $X = G\backslash B\sigma B$ is a closed subset of G, $X \neq G$. We claim that $S_o \subseteq \overline{X}$. So let $a \in S_o$, $a \notin \overline{X}$. Since $0 \in \overline{B} \subseteq \overline{X}$, $a \neq 0$. By Theorem 4.35, G is the disjoint union of $B\theta B (\theta \in W)$. Thus $a \notin \overline{B\theta B}$ for any $\theta \in W$, $\theta \neq \sigma$. By Proposition 6.3, $M = G\overline{B} = \overline{B}G$. Hence $a \in \overline{B}B\sigma B \cap B\sigma B\overline{B} \subseteq \overline{B}\sigma B \cap B\sigma\overline{B}$. Let $\sigma = gT$, $g \in N_G(T)$.

Then there exist $b_1, b_2 \in B$, $u_1, u_2 \in \overline{B}$, such that $a = u_1 g b_1 = b_2 g u_2$. So $u_2 b_1^{-1} = g^{-1} b_2^{-1} u_1 g \in \overline{B} \cap \overline{B}^{-}$. Thus $u_2 b_1^{-1}$ is a diagonal matrix. But $u_2 b_1^{-1} = g^{-1} b_2^{-1} a b_1^{-1} \subset S_o \backslash \{0\}$. This contradicts the assumption that S_o is nil. Thus $S_o \subseteq \overline{X}$. So $S_o \subseteq \overline{X'}$ for some irreducible component X' of X. Since $\dim S_o = p - 1$ and $X' \neq G$, we see that $S_o = \overline{X'}$. This is a contradiction since $S_o \subseteq M \backslash G$. Therefore S_o is not nil. Choose $0 \neq e \in E(S_o)$ such that J_e is maximal in $\mathcal{U}(S_o)$. Let $Y = \{a \mid a \in S_o, a \nmid e\}$. Then Y is a closed set by Lemma 3.25. Let $u \in S_o \backslash Y$. Then $u \mid e$. By Corollary 6.13, $u \in GM_eG$. Hence u is regular. By the maximality of J_e, $u \mathcal{J} e$. Thus $u \in MeM$. Hence $S_o = \overline{MeM} \cup Y$. Since $e \notin Y$ and S_o is irreducible, we see that $S_o = \overline{MeM}$. Now $e \in \Gamma$ for some maximal chain Γ in $E(M)$. Let $B_1 = C_G^r(\Gamma)$, $B_2 = C_G^{\ell}(\Gamma)$. Then B_1, B_2 are Borel subgroups of G by Theorem 7.1. Clearly $B_1 e = eB_1 e$, $eB_2 = eB_2 e$. Then $B_1(eMe)B_2 = eMe$. By Corollary 4.8, $GeMeG$ is closed in M. Since $GeG \subseteq GeMeG$, we see that $S_o = \overline{MeM} = \overline{GeG} \subseteq GeMeG$. Since eMe is regular, we see that every element of S_o is regular. This proves the theorem.

We now proceed as in the author [73] to treat the case when M does not have a zero.

<u>Theorem 7.4</u>. Let M be a connected monoid with group of units G, e a minimal idempotent of M. Then the following conditions are equivalent

(i) M is regular.

(ii) $\overline{\text{rad } G}$ is completely regular.

(iii) G_e is a reductive group.

<u>Proof</u>. (i) $\Rightarrow$ (ii). Let $a \in \overline{\text{rad } G}$. Now $axa = a$ for some $x \in M$. By Propostion 6.3, $x \in \overline{B}$ for some Borel subgroup B of G. Then $\text{rad } G \subseteq B$. So $a \in \overline{B}$. Then a is a regular element of $\overline{B}$. By Corollaries 3.18, 4.12, $a \mathcal{H} e$ for some $e \in E(\overline{B})$.

By Remark 1.3 (iii), $e \in \overline{\text{rad } G}$, $a \mathcal{H} e$ in $\overline{\text{rad } G}$.

(ii) ⇒ (iii). Since e lies in the kernel of M, we see by Theorem 6.30 that $w(e) = 1$, $e \in \overline{\text{rad } G}$. Hence $(\overline{\text{rad } G})_e$ is a completely regular monoid with zero e. By Theorem 5.11, $(\overline{\text{rad } G})_e$ is a torus. Now $e \in \overline{T}$ for some maximal torus T of G. Let B be a Borel subgroup of G with $T \subseteq B$. Now $\text{rad } G_e \subseteq B_1$ for some Borel subgroup B_1 of G, $\text{rad } G_e \triangleleft C_G(e)$. By Lemma 6.28, $u^{-1}B_1u = B$ for some $u \in C_G(e)$. Hence $\text{rad } G_e \subseteq B$ for all $B \in \mathcal{B}(T)$. By Remark 4.39 (ii), $\text{rad } G_e \subseteq T \, \text{rad}_u G$. So $(\text{rad } G_e)_u \subseteq (\text{rad}_u G \subset G_e)^c \subseteq (\overline{\text{rad } G})_e$. But $(\overline{\text{rad } G})_e$ is a torus. Hence $\text{rad } G_e$ is a torus and G_e is a reductive group.

(iii) ⇒ (i). By Theorem 7.3, M_e is regular. Let $a \in M$. Since e lies in the kernel of M, $a \mid e$. By Corollary 6.13, $a \in GM_eG$. So a is regular. This proves the theorem.

We will need the following result of [78] in Chapter 11.

<u>Proposition 7.5</u>. Let M be a connected regular monoid with zero and group of units G. Let T be a maximal torus of G, $\Gamma = \{e_1,\dots,e_k\} \subseteq E(\overline{T})$, $h = e_1 \vee e_2 \vee \dots \vee e_k$. Then $C_G(\Gamma) \subseteq C_G(h)$, $C_G^r(\Gamma) \subseteq C_G^r(h)$ and $C_G^\ell(\Gamma) \subseteq C_G^\ell(h)$.

<u>Proof</u>. First we show that $C_G(\Gamma) \subseteq C_G(h)$. Let T_o denote the radical of the reductive group $C_G(\Gamma)$. Then by Corollary 6.31, $\Gamma \subseteq E(\overline{T}_o)$ and $E(\overline{T}_o)$ is a sublattice of $E(\overline{T})$. Hence $h \in E(\overline{T}_o)$. Since T_o lies in the center of $C_G(\Gamma)$, $C_G(\Gamma) \subseteq C_G(h)$.

We now prove by induction on $|\Gamma| \geq 2$, that $C_G^\ell(\Gamma) \subseteq C_G^\ell(h)$. Let $e = e_1 \vee \dots \vee e_{k-1}$, $f = e_k$. Then $C_G^\ell(e_1,\dots,e_{k-1}) \subseteq C_G^\ell(e)$, $h = e \vee f$. It suffices to show that $C_G^\ell(e,f) \subseteq C_G^\ell(h)$. Let $G' = C_G^\ell(e,f)$, $M' = \overline{G}'$. Let $h' \in E(M')$ such that $h \mathcal{R} h'$. We wish to show that $h = h'$. Let $G'' = C_{G'}^r(e,h)$, $M'' = \overline{G''}$. Clearly $T \subseteq G''$. Since $e \leq h$, we see by Corollary 6.18, that $h' \in M''$. Let

$G''' = C^r_{G''}(f,h)$, $M''' = \overline{G'''}$. Clearly $T \subseteq G'''$. Since $f \le h$, we see that $h' \in M'''$. Now $G''' \subseteq C^{\ell}_G(e,f) \cap C^r_G(e,f) = C_G(e,f) \subseteq C_G(h)$. It follows that $h = h'$. Now let $x \in G'$. Then by Theorem 5.9, there exists $h_1,h_2 \in E(M')$ such that $h \mathcal{R} h_1 \mathcal{L} h_2 \mathcal{R} x^{-1}hx$. Then $h \mathcal{R} h_2$, $h \mathcal{R} xh_2x^{-1}$. By the above, $h_1 = h = xh_2x^{-1}$. So $h \mathcal{L} h_2 = x^{-1}hx$. Hence $hx = hxh$. Thus $G' \subseteq C^{\ell}_G(h)$. Similarly $C^r_G(\Gamma) \subseteq C^r_G(h)$.

The following result is from the author [77].

Proposition 7.6. Let M be a connected regular monoid with zero and groups of units G. Let $e,f \in E(M)$. Then there exists maximal tori T_1,T_2 of G, $e_i,f_i \in E(\overline{T}_i)$, $i = 1,2$ such that $e \mathcal{R} e_1$, $e \mathcal{L} e_2$, $f \mathcal{R} f_1$, $f \mathcal{L} f_2$.

Proof. By Theorem 7.1, $C^r_G(e)$, $C^r_G(f)$, $C^{\ell}_G(e)$, $C^{\ell}_G(f)$ are all parabolic subgroups of G. By Corollary 4.36, there exist maximal tori T_1, T_2 of G such that $T_1 \subseteq C^r_G(e) \cap C^r_G(f)$, $T_2 \subseteq C^r_G(e) \cap C^{\ell}_G(f)$. By Corollary 6.10, there exists $x_1,x_2 \in C^r_G(e)$, $y_1 \in C^r_G(f)$, $y_2 \in C^{\ell}_G(f)$ such that $e_1 = x_1^{-1}ex_1$, $f_1 = y_1^{-1}fy_1 \in E(\overline{T}_1)$, $e_2 = x_2^{-1}ex_2$, $f_2 = y_2^{-1}fy_2 \in E(\overline{T}_2)$. The result follows.

Definition 7.7. Let S be a regular semigroup. Then,

(i) $S/\mathcal{R}$ is the partially ordered set $\{eS \mid e \in E(S)\}$ under inclusion, $S/\mathcal{L}$ is the partially ordered set $\{Se \mid e \in E(S)\}$ under inclusion.

(ii) If $e,f \in E(S)$, then the sandwich set, $\text{sand}(e,f) = \{h \mid h \in fSe,\ h$ is an inverse of $ef\}$.

Remark 7.8. (i) The sandwich set sand (e,f) plays a crucial role in the theory of regular semigroups and biordered sets. See [52], Chapter 12. For regular semigroups, sand(e,f) is always non–empty and a rectangular band (i.e., an idempotent semigroup satisfying the identity $xyz = xz$).

(ii) The partially ordered sets $S/\mathcal{R}$, $S/\mathcal{L}$ are the starting point of Grillet's theory of regular semigroups [27], [28].

The following result is due to the author [77].

Theorem 7.9. Let M be a connected regular monoid with group of units G. Then

(i) $M/\mathcal{R}$, $M/\mathcal{L}$ are relatively complemented, complete lattices with all maximal chains having the same finite length $\mathrm{ht}(M)$.

(ii) If $e,f \in E(M)$, then $\mathrm{sand}(e,f) = \{h \mid h = e'f'$ for some $e',f' \in E(\overline{T})$, T a maximal torus of G, $e \,\mathcal{L}\, e'$, $f \,\mathcal{R}\, f'\} = \{h \mid h = e'f' = f'e'$ for some $e',f' \in E(M)$ with $e \,\mathcal{L}\, e'$, $f \,\mathcal{R}\, f'\}$.

(iii) If $e \in E(M)$, $a \in M$, then $a \,\mu\, e$ if and only if $a \in C(H)$ where H is the $\mathcal{H}$-class of e.

Proof. (i) That all maximal chains in $M/\mathcal{R}$ have length $\mathrm{ht}(M)$ follows from Theorem 6.20. Since M is the maximum element of $M/\mathcal{R}$, it suffices to show that $M/\mathcal{R}$ is a $\wedge$-semilattice which is relatively complemented. Let $eM, fM \in M/\mathcal{R}$ where $e,f \in E(M)$. By Proposition 7.6, there exists a maximal torus T of G, $e_1,f_1 \in E(\overline{T})$ such that $eM = e_1M$, $fM = f_1M$. Then, clearly $eM \cap fM = eM \wedge fM = e_1f_1M$. Thus $M/\mathcal{R}$ is a complete lattice. Next let $e,f,h \in E(M)$ such that $eM \supset fM \supset hM$. Without loss of generality, we may assume that $e > f > h$. By Corollary 6.10, there exists a maximal torus T of G such that $e,f,h \in E(\overline{T})$. But $E(\overline{T})$ is relatively complemented by Theorem 6.20. So there exists $f' \in E(\overline{T})$, $e > f' > h$ such that $ff' = h$. Then $eM \supset f'M \supset hM$, $fM \cap f'M = fM \wedge f'M = hM$. Hence $M/\mathcal{R}$ is relatively complemented.

(ii) By Proposition 7.6, there exists a maximal torus T of G, $e_1,f_1 \in E(\overline{T})$ such that $e \,\mathcal{L}\, e_1$, $f \,\mathcal{R}\, f_1$. By Corollary 6.19, $C_G^{\ell}(e) = C_G^{\ell}(e_1)$, $C_G^{r}(f) = C_G^{r}(f_1)$. So by Theorem 6.16, $G_1 = G_G^{\ell}(e) \cap C_G^{r}(f)$ is a connected group and $fMe = f_1Me_1 \subseteq \overline{G_1}$. Clearly $h = e_1f_1 = f_1e_1 \in \mathrm{sand}(e,f)$ and $x^{-1}hx \in \mathrm{sand}(e,f)$ for all $x \in G_1$. Let $h' \in$ sand(e,f). Then $h' \in fMe$ and by Remark 7.8 (i), $h \mathcal{J} h'$ in fMe. By Corollary 6.8, $h' = x^{-1}hx$ for some $x \in G_1$. Thus sand(e,f) = $\{x^{-1} hx \mid x \in G_1\}$ and the result follows.

(iii) By Remark 1.21 (ii), we are reduced to the case when $e = 1$. Clearly $a \mu 1$ if and only if $a \in C_G(E(M))$. By Theorem 7.1, $C_G(E(M))$ is the intersection of all Borel subgroups of G, which is C(G) by Theorem 4.32.

<u>Remark 7.10</u>. (i) Let $J/\mathcal{R} = \{eM \mid e \in E(J)\}$. Then $M/\mathcal{R} = \bigcup_{J \in \mathcal{U}(M)} J/\mathcal{R}$. Let $e \in E(J)$. Then $E(J) = \{xex^{-1} \mid x \in G\}$. Let $x_1, x_2 \in G$, $f_i = x_i e x_i^{-1}$, $i = 1,2$. Then $f_1M = f_2M$ if and only if $x_1C_G^r(e) = x_2C_G^r(e)$. Thus $J/\mathcal{R}$ is in 1–1 correspondence with $G/C_G^r(e)$. But $C_G^r(e)$ is a parabolic subgroup of G by Theorem 7.1. Hence $G/C_G^r(e)$ is a projective variety by Theorem 4.7. Thus $M/\mathcal{R}$ can be thought of as a lattice ordered projective variety.

(ii) Theorem 7.9 (iii) shows that M is a central extension of M/μ and hence susceptible to a cohomological approach of [42], [43]. M/μ will be studied further in Chapters 14, 15.

In the local study of connected regular monoids, Green's relations represent the first step. The ultimate goal is the study of conjugacy classes. For $\mathcal{M}_n(K)$, Green's relations amount to row and column equivalence. Studying conjugacy classes yields the Jordan canonical forms. For algebraic monoids, the problem is much more difficult. Some advances have been recently made by the author [84]. We will describe the main results of [84] without proofs.

Let M be a connected regular monoid with zero and group of units G. Let T be a maximal torus of G, $W = N_G(T)/T$. Let $e \in E(T)$, $\sigma = nT \in W$. Let

$$M_{e,\sigma} = eC_G(e^\theta \mid \theta \in <\sigma>)\sigma$$

Let $V = C_G(e^\theta \mid \theta \in <\sigma>)$, $Y = \{g \in V \mid ge = e\}$. Let

$$H = \prod_{\theta \in \langle \sigma \rangle} Y^{\theta}, \; G_{e,\sigma} = V/H$$

For $x \in V$, let $x^* = nx^{-1}n^{-1} \in V$. Clearly $H^* = H$. So $*$ induces an anti-automorphism $*$ on $G_{e,\sigma}$. Define $\xi = \xi_{e,\sigma}: M_{e,\sigma} \twoheadrightarrow G_{e,\sigma}$ as follows. If $a = evn \in M_{e,\sigma}$, $v \in V$, then

$$\xi(a) = vH \in G_{e,\sigma}$$

The main theorem of [84] is the following.

Theorem 7.11. Every element of M is conjugate to an element of some $M_{e,\sigma}$. Let $a,b \in M_{e,\sigma}$. Then a is conjugate to b in M if and only if there exists $x \in G_{e,\sigma}$ such that $x\xi(a)x^* = \xi(b)$.

If $F = \langle 1, e^{\theta} \mid \theta \in \langle \sigma \rangle \rangle$, then $M_{e,\sigma} \subseteq FN_G(F)$ and $FN_G(F)$ is an inverse semigroup. Thus

Corollary 7.12. M is a union of its inverse submonoids.

Remark 7.13. That $\mathcal{M}_n(K)$ is a union of its inverse submonoids was noted by Schein [104].

Let $e \in E(\overline{T})$, $\sigma \in W$. Then for $k \in \mathbb{Z}^+$, $M^k_{e,\sigma} = 0$ if and only if $(e\sigma)^k = 0$. For $\mathcal{M}_n(K)$, the groups $G_{e,\sigma}$ are all trivial for nilpotent $e\sigma$.

Example 7.14. Let $G_0 = (A \otimes (A^{-1})^t \mid A \in SL(3,K))$, $G = K^*G_0$, $M = \overline{KG_0}$. Let $S = M \backslash G$. Then

$$E(S) = \{e \otimes f \mid e^2 = e, f^2 = f \in \mathcal{M}_3(K), ef^t = f^te = 0\}.$$

In particular

$$e = \begin{bmatrix} 1&0&0 \\ 0&0&0 \\ 0&0&0 \end{bmatrix} \otimes \begin{bmatrix} 0&0&0 \\ 0&1&0 \\ 0&0&1 \end{bmatrix}, \ f = \begin{bmatrix} 0&0&0 \\ 0&1&0 \\ 0&0&0 \end{bmatrix} \otimes \begin{bmatrix} 1&0&0 \\ 0&0&0 \\ 0&0&1 \end{bmatrix} \in E(M)$$

Also if $\sigma = \begin{bmatrix} 0&1&0 \\ 1&0&0 \\ 0&0&-1 \end{bmatrix} \otimes \begin{bmatrix} 0&1&0 \\ 1&0&0 \\ 0&0&-1 \end{bmatrix} \in W(G)$, then $e^{\sigma} = f$ and $(e\sigma)^2 = 0$. So $(M_{e,\sigma})^2 = 0$. The group $G_{e,\sigma}$ can be seen to be the one dimensional torus with * being given by: $x \to x^{-1}$. Thus by Theorem 7.11, the number of conjugacy classes of nilpotent elements of M is infinite. However if C denotes the center of G, then the number of conjugacy classes of nilpotent elements in M/C is finite.

Example 7.15. Suppose char $K \neq 2$, $n \in \mathbb{Z}^+$, $n \geq 2$. For $r \in \mathbb{Z}^+$, let J_r denote the $r \times r$ matrix $\begin{bmatrix} & & 1 \\ & \cdot^{\cdot^{\cdot}} & \\ 1 & & \end{bmatrix}$. Let G_o consist of all $A \in SL(2n+1, K)$ such that

$$A^t \begin{bmatrix} 1 & 0 \\ 0 & J_{2n} \end{bmatrix} A = \begin{bmatrix} 1 & 0 \\ 0 & J_{2n} \end{bmatrix}$$

Thus [34; Section 7.2], G_o is the special orthogonal group of type B_n. Let $G = K^*G_o$, $M = \overline{KG_o}$. Then

$$e = \begin{bmatrix} 0&0&0 \\ 0&I_n&0 \\ 0&0&0 \end{bmatrix}, \ f = \begin{bmatrix} 0&0&0 \\ 0&0&0 \\ 0&0&I_n \end{bmatrix} \in E(M).$$

If $\sigma = \begin{bmatrix} \pm 1 & 0 \\ 0 & J_{2n} \end{bmatrix} \in W(G)$, then $e^{\sigma} = f$ and $(e\sigma)^2 = 0$. Thus $M_{e,\sigma}^2 = 0$. It can be seen that $G_{e,\sigma} \cong PGL(n,K)$ with the anti-automorphism * on $G_{e,\sigma}$ given by: $A \to J_n A^t J_n$. Thus the conjugacy classes of elements of $M_{e,\sigma}$ in M are in one to one correspondence with the congruence classes of $PGL(n,K)$. Two elements $A_1, A_2 \in$

PGL(n,K) are congruent if $C^t A_1 C = A_2$ for some $C \in PGL(n,K)$.

Contrast the situation for monoids with that for reductive groups, where the number of conjugacy classes of unipotent elements is always finite [44].

8 DIAGONAL MONOIDS

By a connected diagonal monoid, we mean a connected monoid $\bar{T}$ such that the group of units T is torus. The importance of knowing more about $E(\bar{T})$ is clear from the previous chapters. The author [66] was pleasantly surprised to discover that $E(\bar{T})$ is just the face lattice of a rational polytope. However, it should be noted that the connection between diagonal monoids and rational polyhedral cones is already clear from the theory of torus embeddings [37], [58]. Our needs are more elementary and we will follow [66].

If $X \subseteq \mathbb{R}^n$, then conv X will denote the convex hull of X. The convex hull of a finite set in $\mathbb{R}^n$ is called a polytope. If P is a polytope, $F \subseteq P$, then F is a face of P, if for all $a,b \in P$, $\alpha \in (0,1)$, $\alpha a + (1-\alpha)b \in F$ if and only if $a,b \in F$. The set of all faces of P forms a finite lattice $\mathcal{F}(P)$, called the face lattice of P. The singleton faces are called vertices. Then P is the convex hull of its vertices. If the vertices are all rational, then P is said to be a rational polytope. We refer to [30] for details.

Let M be a connected monoid with zero 0, group of units G. By a character of M we mean a homomorphism $\chi: M \to (K,\cdot)$ such that $\chi(1) = 1$, $\chi(0) = 0$. Let $\mathcal{X}(M)$ denote the commutative semigroup of all characters on M. If $\chi \in \mathcal{X}(M)$, let $\bar{\chi} \in \mathcal{X}(G)$ denote the restriction of χ tp G. Since $M = \bar{G}$, the homomorphism $\chi \to \bar{\chi}$ is injective. Thus

$$\mathcal{X}(M) \hookrightarrow \mathcal{X}(G) \qquad (6)$$

By Remark 4.18 (i), $\mathcal{X}(G)$ is linearly independent in the vector space of all K–valued functions on G. By (6), we have,

Lemma 8.1. $\mathcal{X}(M)$ is linearly independent in the vector space of all K–valued functions on M.

Let T be a closed connected subgroup of $\mathcal{D}_n^*(K)$ such that $0 \in \bar{T}$ in $\mathcal{D}_n(K)$. If $a = \text{diag}(\alpha_1,\ldots,\alpha_n) \in \mathcal{D}_n(K)$, $f \in K[x_1,\ldots,x_n]$, then let $f(a) = f(\alpha_1,\ldots,\alpha_n) \in K$.

Corollary 8.2. Let $a \in \mathcal{D}_n(K)$, $a \notin \bar{T}$. Then there exist monomials $f,g \in K[x_1,\ldots,x_n]$ such that $f(b) = g(b)$ for all $b \in \bar{T}$, $f(a) \neq g(a)$.

Proof. There exists $p \in K[x_1,\ldots,x_n]$ such that $p(b) = 0$ for all $b \in \bar{T}$, $p(a) \neq 0$. Since $0 \in \bar{T}$, the constant term of p is zero. So $0 \neq p = \sum_{i=1}^{m} \alpha_i p_i, \alpha_i \in K$, $p_1,\ldots,p_m$ monomials in $K[x_1,\ldots,x_n]$, $m \geq 2$. Chose p such that m is minimal. Now each p_i, restricted to $\bar{T}$ is in $\mathcal{X}(\bar{T})$. Hence by Lemma 8.1, $p_i = p_j$ on $\bar{T}$ for some $i \neq j$. We claim that $p_i(a) \neq p_j(a)$. For suppose $p_i(a) = p_j(a)$. Let $q \in K[x_1,\ldots,x_n]$ be obtained by replacing the monomial p_i by p_j in p. Then $q(b) = 0$ for all $b \in \bar{T}$ and $q(a) = p(a) \neq 0$. This contradicts the minimality of m. Hence $p_i(a) \neq p_j(a)$.

Definition 8.3. Let $u_1,\ldots,u_n \in \mathbb{R}^m$, $u = (u_1,\ldots,u_n)$, $P = \text{conv}\{u_1,\ldots,u_n\}$. If $F \in \mathcal{F}(P)$, then let $e_F = \text{diag}(e_1,\ldots,e_n)$ where $e_i = 1$ if $u_i \in F$, $e_i = 0$ if $u_i \notin F$. Let $E(u) = \{e_F \mid F \in \mathcal{F}(P)\} \cong \mathcal{F}(P)$.

If $u = (i_1,\ldots,i_m) \in \mathbb{Z}^m$, $a = \text{diag}(a_1,\ldots,a_m) \in \mathcal{D}_m^*(K)$, then we let $u(a) = a_1^{i_1} \cdots a_m^{i_m} \in K^*$.

<u>Theorem 8.4</u>. Let $u_1,\ldots,u_n \in \mathbb{Z}^m$, $u = (u_1,\ldots,u_n)$. Let $T_o = \{\text{diag}(u_1(a),\ldots,u_n(a)) \mid a \in \mathscr{D}_m^*(a)\} \subseteq \mathscr{D}_n^*(K)$. Let $T = K^*T_o$, $\bar{T}$ the closure in $\mathscr{D}_n(K)$. Then $E(\bar{T}) = E(u)$.

<u>Proof</u>. Let $F \in \mathscr{F}(P)$. Suppose $e_F \notin E(\bar{T})$. Then by Corollary 8.2, there exist monomials $f,g \in K[x_1,\ldots,x_n]$ such that $f(b) = g(b)$ for all $b \in T$, $f(e_F) \neq g(e_F)$. Now $f(e_F), g(e_F) \in \{0,1\}$. So we may assume that

$$f(e_F) = 1, \; g(e_F) = 0 \tag{7}$$

Let $f = x_1^{i_1} \ldots x_n^{i_n}$, $g = x_1^{j_1} \ldots x_n^{j_n}$. Thus $i_1,\ldots,i_n, j_1,\ldots,j_n$ are non–negative integers, $i = i_1 + \ldots + i_n$, $j = j_1 + \ldots + j_n \in \mathbb{Z}^+$. Now for all $\alpha \in K$, $\alpha^i = f(\alpha \cdot 1) = g(\alpha \cdot 1) = \alpha^j$. Since K is infinite, $i = j$. Now for all $a \in \mathscr{D}_m^*(K)$

$$u_1(a)^{i_1} \ldots u_n(a)^{i_n} = u_1(a)^{j_1} \ldots u_n(a)^{j_n}$$

Since K is infinite, we see that

$$\sum_{k=1}^{n} i_k u_k = \sum_{k=1}^{n} j_k u_k$$

Let $i_k' = i_k/i$, $j_k' = j_k/j$. Since $i = j$,

$$\sum_{k=1}^{n} i_k' u_k = \sum_{k=1}^{n} j_k' u_k, \quad \sum_{k=1}^{n} i_k' = \sum_{k=1}^{n} j_k' = 1.$$

Now let $e_F = \mathrm{diag}(e_1,\ldots,e_n)$. Then by (7), $i_k = 0$ for any $u_k \notin F$. Hence $\sum_{k=1}^{n} i_k' u_k \in F$. Also by (7), $j_\ell \neq 0$ for at least one $u_\ell \notin F$. Hence $\sum_{k=1}^{n} j_k' u_k \notin F$. This contradiction shows that $e_F \in \bar{T}$.

Now let $e = \mathrm{diag}(e_1,\ldots,e_n) \in E(\bar{T})$. Let $\Omega = \{k \mid e_k = 1\}$, $F = \mathrm{conv}\{u_k \mid k \in \Omega\}$. We claim that $F \in \mathscr{F}(P)$, $e = e_F$. Let $\emptyset \neq \Gamma \subseteq \{1,\ldots,n\}$. Suppose that there exist $\varepsilon_\ell \in \mathbb{R}^+ (\ell \in \Gamma)$ such that $\sum_{k\in\Gamma} \varepsilon_k u_k \in F$, $\sum_{\ell\in\Gamma} \varepsilon_\ell = 1$. It suffices to show that then $\Gamma \subseteq \Omega$. Now there exists $\emptyset \neq \Omega' \subseteq \Omega$, $\delta_p \in \mathbb{R}^+$ $(p \in \Omega')$ such that

$$\sum_{\ell\in\Gamma} \varepsilon_\ell u_\ell = \sum_{p\in\Omega'} \delta_p u_p,\ \sum_{\ell\in\Gamma} \varepsilon_\ell = \sum_{p\in\Omega'} \delta_p \tag{8}$$

Since $u_1,\ldots,u_n \in \mathbb{Z}^m$, we see by (8), that $\varepsilon_\ell, \delta_p (\ell \in \Gamma, p \in \Omega')$ represent a solution to a suitable homogeneous system of linear equations with integer coefficients. But the solution space has a rational basis. Thus we can find $i_\ell, j_p \in \mathbb{Z}^+$ $(\ell \in \Gamma, p \in \Omega')$ such that

$$\sum_{\ell\in\Gamma} i_\ell u_\ell = \sum_{p\in\Omega'} j_p u_p,\ \sum_{\ell\in\Gamma} i_\ell = \sum_{p\in\Omega'} j_p$$

Thus

$$\prod_{\ell\in\Gamma} (\alpha u_\ell(a))^{i_\ell} = \prod_{p\in\Omega'} (\alpha u_p(a))^{j_p}$$

for all $a \in \mathscr{D}_m^*(K)$, $\alpha \in K$. Since $\bar{T} = \overline{KT_o}$, we see that for all $b = \mathrm{diag}(b_1,\ldots,b_n) \in \bar{T}$,

$$\prod_{\ell\in\Gamma} b_\ell^{i_\ell} = \prod_{p\in\Omega'} b_p^{j_p}.$$

Since $e = \mathrm{diag}(e_1,\dots,e_n) \in \overline{T}$,

$$\prod_{\ell\in\Gamma} e_\ell = \prod_{p\in\Omega'} e_p = 1$$

Thus $\Gamma \subseteq \Omega$, proving the theorem.

The situation in Theorem 8.4 arises as follows. Let $G_o \subseteq GL(n,K)$ be a reductive group, T_o a maximal torus of G_o, $T_o \subseteq \mathscr{D}_n^*(K)$. Let $T = K^*T_o$, $M = \overline{KG_o}$, $G = K^*G_o$. Then by Theorem 7.3, M is a connected regular monoid with zero. So by Proposition 6.1, Corollary 6.10, $M = GE(\overline{T})G$, $E(M) = \bigcup_{x\in G} x^{-1}E(\overline{T})x$. Thus our interest in $E(\overline{T})$. Let $\dim T_o = m$. Then $T_o \cong \mathscr{D}_m^*(K)$, $\mathscr{X}(\mathscr{D}_m^*(K)) \cong \mathbb{Z}^m$ by Theorem 4.19. So by Remark 4.18 (ii), there exist $u_1,\dots,u_n \in \mathbb{Z}^m$ such that $T_o = \{\mathrm{diag}(u_1(a),\dots,u_n(a)) \mid a \in \mathscr{D}_m^*(K)\}$. By Theorem 8.4, $E(\overline{T}) = E(u_1,\dots,u_n)$.

Example 8.5. Let $G_o = \{A \otimes (A^{-1})^t \mid A \in SL(3,K)\}$, $M = \overline{KG_o} \subseteq \mathscr{M}_9(K)$. We wish to compute $E(M)$. Now $T_o = \{A \otimes A^{-1} \mid A \in \mathscr{D}_3^*(K), \det A = 1\}$ is a maximal torus of G_o. Let $T = K^*T_o$, $G = K^*G_o$. Clearly

$$\begin{aligned} T_o &= \{\mathrm{diag}(a,b,1/ab) \otimes \mathrm{diag}(1/a,1/b,ab) \mid a,b \in K^*\} \\ &= \{\mathrm{diag}(1,a/b,a^2b,b/a,1,ab^2,1/a^2b,1/ab^2,1) \mid a,b \in K^*\}. \end{aligned}$$

Correspondingly let $u_1 = (0,0)$, $u_2 = (1,-1)$, $u_3 = (2,1)$, $u_4 = (-1,1)$, $u_5 = (0,0)$, $u_6 = (1,2)$, $u_7 = (-2,-1)$, $u_8 = (-1,-2)$, $u_9 = (0,0)$. Let $P = \mathrm{conv}\{u_1,\dots,u_9\}$, $u = (u_1,\dots,u_9)$. Then P is a hexagon with vertices $(2,1)$, $(1,2)$, $(-1,1)$, $(-2,-1)$, $(-1,-2)$, $(1,-1)$. Let $e_1 = \mathrm{diag}(1,0,0)$, $e_2 = \mathrm{diag}(0,1,0)$, $e_3 = \mathrm{diag}(0,0,1)$, $f_1 = e_2 + e_3$, $f_2 = e_1 + e_3$, $f_3 = e_1 + e_2$. It is easily verified that $E(\overline{T}) = E(u) = \{0,1\} \cup \{e_i \otimes e_j \mid i,j = 1,2,3,\ i \neq j\} \cup \{e_i \otimes f_i \mid i = 1,2,3\} \cup \{f_i \otimes e_i \mid i = 1,2,3\}$. Now an idempotent in M is of the form $A^{-1}hA \otimes A^t h'(A^{-1})^t$ where $h \otimes h' \in E(\overline{T})$,

$A \in SL(3,K)$. It follows that $\mathscr{U}(M) = \{G,J_1,J_2,J_0,0\}$ with $J_1 > J_0$, $J_2 > J_0$. Also $E(J_1) = \{e \otimes f \mid e^2 = e,\ f^2 = f \in \mathscr{M}_3(K),\ \rho(e) = 1,\ \rho(f) = 2,\ ef^t = f^te = 0\}$, $E(J_2) = \{e \otimes f \mid e^2 = e,\ f^2 = f \in \mathscr{M}_3(K),\ \rho(e) = 2,\ \rho(f) = 1,\ ef^t = f^te = 0\}$, $E(J_0) = \{e \otimes f \mid e^2 = e,\ f^2 = f \in \mathscr{M}_3(K),\ ef^t = f^te = 0\}$.

Example 8.6. Let $G_0 = \{A \oplus (A^{-1})^t \mid A \in SL(3,K)\}$, $M = \overline{KG_0} \subseteq \mathscr{M}_6(K)$. Now $T_0 = \{A \oplus A^{-1} \mid A \in \mathscr{D}_3^*(K),\ \det A = 1\}$ is a maximal torus of G_0. Let $T = K^*T_0$, $G = K^*G_0$. Clearly

$$T_0 = \{\mathrm{diag}(a,b,1/ab,1/a,1/b,ab) \mid a,b \in K^*\}$$

Correspondingly let $u_1 = (1,0)$, $u_2 = (0,1)$, $u_3 = (-1,-1)$, $u_4 = (-1,0)$, $u_5 = (0,-1)$, $u_6 = (1,1)$. Let $P = \mathrm{conv}\{u_1,\ldots,u_6\}$, $u = (u_1,\ldots,u_6)$. Then P is a hexagon with vertices $u_1,\ldots,u_6$. Let $e_1 = \mathrm{diag}(1,0,0)$, $e_2 = \mathrm{diag}(0,1,0)$, $e_3 = \mathrm{diag}(0,0,1)$. Then $E(\overline{T}) = E(u) = \{0,1\} \cup \{e_i \oplus e_j \mid i,j = 1,2,3,\ i \neq j\} \cup \{e_i \oplus 0 \mid i = 1,2,3\} \cup \{0 \oplus e_i \mid i = 1,2,3\}$. Any idempotent of M is of the form $A^{-1}hA \oplus A^th'(A^{-1})^t$ where $h \oplus h' \in E(\overline{T})$. Hence $E(M)\backslash\{1\} = \{e \oplus f \mid e^2 = e,\ f^2 = f \in \mathscr{M}_3(K),\ \rho(e) \leq 1,\ \rho(f) \leq 1\}$.

We now wish to prove the converse of Theorem 8.4. So let T be a closed connected subgroup of $\mathscr{D}_n^*(K)$ such that $0 \in \overline{T}$ in $\mathscr{D}_n(K)$. Let $\dim T = m$. Then by Theorem 4.19, there exists $v_1,\ldots,v_n \in \mathbb{Z}^m$ such that $T = \{\mathrm{diag}(v_1(a),\ldots,v_n(a)) \mid a \in \mathscr{D}_m^*(K)\}$. Since $0 \in \overline{T}$, we see that $0 \notin \langle v_1,\ldots,v_n \rangle$. Thus $0 \notin \mathrm{conv}\{v_1,\ldots,v_n\}$. So it is easy to see [27; Theorem 2.21] that there exists $v \in \mathbb{R}^m$ such that the inner product $v \cdot v_i > 0$, $i = 1,\ldots,n$. Since $v_1,\ldots,v_n \in \mathbb{Z}^m$ we can then choose $v \in \mathbb{Z}^m$. We can find $v' \in \mathbb{Q}^m$ such that $v \cdot v' = 1$. Let $u_i = (v_i/v \cdot v_i) - v'$. Then u_i is a rational point in $\{x \in \mathbb{R}^m \mid v \cdot x = 0\} \cong \mathbb{R}^{m-1}$. Let $u = (u_1,\ldots,u_n)$. It is routinely verified as in the proof of Theorem 8.4 that $E(\overline{T}) = E(u)$. Hence we have,

Theorem 8.7. Let T be a closed connected subgroup of $\mathscr{D}_n^*(K)$ such that $0 \in \overline{T}$ in $\mathscr{D}_n(K)$. Let $\dim T = m$. Then there exists $u_1,\dots,u_n \in \mathbb{Z}^{m-1}$ such that with $u = (u_1,\dots,u_n)$, $P = \mathrm{conv}\{u_1,\dots,u_n\}$, $E(\overline{T}) = E(u) \cong \mathscr{F}(P)$.

Remark 8.8. If $\dim T = 2$, then P is of course a line and hence $|E(\overline{T})| = 4$.

We will need the next result in Chapter 10.

Corollary 8.9. Let $T_o \subseteq \mathscr{D}_n^*(K)$ be a torus of dimension 1 containing non-scalar matrices. Let $T = K^*T_o$, $E(\overline{T}) = \{1,e,f,0\}$. Let $T_1 = \{t \otimes t^{-1} \mid t \in T_o\} \subseteq \mathscr{D}_{n^2}(K)$, $T_2 = K^*T_1$. Then $E(\overline{T}_2) = \{1, e \otimes f, f \otimes e, 0\}$.

Proof. There exist $i_1,\dots,i_n \in \mathbb{Z}$, not all equal such that $T_o = \{\mathrm{diag}(a^{i_1},\dots,a^{i_n}) \mid a \in K^*\}$. Let $\alpha = \max\{i_1,\dots,i_n\}$, $\gamma = \min\{i_1,\dots,i_n\}$. Let $e = \mathrm{diag}(e_1,\dots,e_n)$ where $e_r = 1$ if $i_r = \alpha$, 0 otherwise. Let $f = \mathrm{diag}(f_1,\dots,f_n)$ where $f_r = 1$ if $i_r = \gamma$, 0 otherwise. By Theorem 8.4, $E(\overline{T}) = \{1,e,f,0\}$. Now $T_1 = \{\mathrm{diag}(\dots,a^{i_k - i_\ell},\dots) \mid a \in K^*\}$, $\alpha - \gamma = \max\{i_k - i_\ell \mid k,\ell = 1,\dots,n\}$, $\gamma - \alpha = \min\{i_k - i_\ell \mid k,\ell = 1,\dots,n\}$. It follows from Theorem 8.4 that $E(\overline{T}_2) = \{1, e \otimes f, f \otimes e, 0\}$.

Let the height function ht be as in Definition 6.21.

Corollary 8.10. Let T be a connected diagonal monoid with zero 0. Let $e,e' \in E(T)$ such that $\mathrm{ht}(e) = \mathrm{ht}(e') = p > 0$. Then there exists $e = e_o, e_1,\dots,e_k = e'$, $f_1,\dots,f_k \in E(T)$ such that $\mathrm{ht}(e_i) = p$, $\mathrm{ht}(f_i) = p-1$, $e_i > f_i$, $e_{i-1} > f_i$, $i = 1,\dots,k$.

Proof. Let $\dim T = m$. Proceeding by induction on m we are easily reduced to the case when $\mathrm{ht}(e) = \mathrm{ht}(e') = m-1$. By Theorem 8.7, $E(T) \cong \mathscr{F}(P)$ for some polytope P. Then $E(T)$ is anti-isomorphic to $\mathscr{F}(P^*)$ where P^* is the dual polytope of P (see [30; Section 3.4]). Hence the elements of $E(T)$ of height $m-1$ correspond to the vertices of P^*. By a result of Balinski [2] (or see [30; Section 11.3]), any two vertices of a polytope are connected by a sequence of edges. The result follows.

By Corollaries 6.10, 6.22, 8.10, we have,

Corollary 8.11. Let M be a connected monoid, $J, J' \in \mathcal{U}(M)$, $ht(J) = ht(J') = p > 0$. Then there exists $J = J_o, J_1,\ldots,J_k = J', J_1^*,\ldots,J_k^* \in \mathcal{U}(M)$ such that $ht(J_i) = p$, $ht(J_i^*) = p - 1$, $J_i > J_i^*$, $J_{i-1} > J_i^*$, $i = 1,\ldots,k$.

Corollary 8.12. Let T be a connected diagonal monoid with zero. Let $\emptyset \neq \Gamma \subseteq E(\overline{T})$ such that $(\Gamma,\leq)$ is a relatively complemented lattice with all maximal chains having length equal to $\dim T$. Then $\Gamma = E(\overline{T})$.

Proof. Let $\dim T = m$. We prove the result by induction on m. Let $X = \{e \in E(\overline{T}) \mid ht(e) = m - 1\}$. If $e \in X \cap \Gamma$, then $\Gamma' = \overline{eT} \cap \Gamma$ satisfies the hypothesis with respect to $\overline{eT}$. Hence $E(\overline{eT}) = \Gamma' \subseteq \Gamma$. Thus it suffices to show that $X \subseteq \Gamma$. Suppose not. Then by Corollary 8.10, there exists $e_1 \in X \cap \Gamma$, $e_2 \in X\backslash\Gamma$ such that with $f = e_1e_2$, $ht(e_1) = ht(e_2) = m - 1$, $ht(f) = m - 2$. By the above $f \in \Gamma$. Since Γ is relatively complemented there exists $e_1' \in \Gamma \cap X$ such that $e_1e_1' = f$. Hence $1,e_1,e_1',e_2,f$ are distinct elements of $E(\overline{T}_f)$. By Theorem 6.20, $\dim T_f = 2$. By Remark 8.8, $|E(\overline{T}_f)| = 4$. This contradiction completes the proof.

Corollary 8.13. Let T be a connected diagonal monoid with zero. Let σ be an automorphism of $E(\overline{T})$ fixing a maximal chain Γ in $E(\overline{T})$. Then $\sigma = 1$.

Proof. We prove by induction on $\dim T = m$. Let $\Omega = \{e \in E(\overline{T}) \mid e^\sigma = e\}$, $X = \{e \in E(\overline{T}) \mid ht(e) = m - 1\}$, $\Gamma \cap X = \{f\}$. Then $\Gamma' = \Gamma\backslash\{1\}$ is a maximal chain of $E(f\overline{T})$. Since $f^\sigma = f$, σ restricts to an automorphism of $E(f\overline{T})$. By induction hypothesis, $E(f\overline{T}) \subseteq \Omega$. Suppose $\Omega \neq E(\overline{T})$. Then by Corollary 8.10, there exist

$e_1, e_2 \in X$ such that $E(e_1T) \subseteq \Omega$, $E(e_2T) \nsubseteq \Omega$, $ht(e_3) = m - 2$ where $e_3 = e_1e_2$. Then $e_3 \in \Omega$. By Remark 8.8, $E(T_{e_3}) = \{1, e_1, e_2, e_3\}$. Since $e_1^\sigma = e_1$ and $e_3^\sigma = e_3$, we see that $e_2^\sigma = e_2$. So $e_2 \in \Omega$. Extend $\{e_2 > e_3\}$ to a maximal chain Λ of $E(e_2T)$. Then $\Lambda \backslash \{e_2\} \subseteq E(e_3T) \subseteq E(e_1T) \subseteq \Omega$. Hence $\Lambda \subseteq \Omega$. By induction hypothesis, $E(eT) \subseteq \Omega$, a contradiction. Hence $\Omega = E(T)$, completing the proof.

Let $T_1 \subseteq \mathscr{D}_m^*(K)$, $T_2 \subseteq \mathscr{D}_n^*(K)$ be tori such that $0 \in \overline{T_1}$ in $\mathscr{D}_m(K)$, $0 \in \overline{T_2}$ in $\mathscr{D}_n(K)$. Let $\phi: \overline{T_1} \to \overline{T_2}$ be a homomorphism such that $\phi(1) = 1$, $\phi(0) = 0$. If $\chi \in \mathscr{X}(\overline{T_2})$, let $\overline{\phi}(\chi) \in \mathscr{X}(\overline{T_1})$ be given by $\overline{\phi}(\chi)(a) = \chi(\phi(a))$. Thus $\overline{\phi}: \mathscr{X}(\overline{T}) \to \mathscr{X}(\overline{T_1})$ is a homomorphism. Next, let $\psi: \mathscr{X}(\overline{T_2}) \to \mathscr{X}(\overline{T_1})$ be a homomorphism. Let $\chi_1, \ldots, \chi_n \in \mathscr{X}(\overline{T_2})$ denote the n projections of $\overline{T_2}$ into K. Let $\hat{\psi}: \overline{T_1} \to \overline{T_2}$ be given by $\hat{\psi}(a) = (\psi(\chi_1)(a), \psi(\chi_2)(a), \ldots, \psi(\chi_n)(a))$. Then $\hat{\psi}$ is a homomorphism, $\hat{\psi}(1) = 1$, $\hat{\psi}(0) = 0$.

Let $\overline{T}$ be a connected diagonal monoid with zero, $\dim T = n$. So by (6), Lemma 8.1, $\mathscr{X}(\overline{T})$ is a finitely generated subsemigroup of $(\mathbb{Z}^n, +)$, $0 \notin \mathscr{X}(\overline{T})$. So $\mathscr{X}(\overline{T})$ is <u>totally cancellative</u>, i.e. it is cancellative and for all $a, b \in \mathscr{X}(\overline{T})$, $k \in \mathbb{Z}^+$, $ka = kb$ implies $a = b$. Conversely let $u_1, \ldots, u_m \in \mathbb{Z}^n$, $0 \notin \langle u_1, \ldots, u_n \rangle$. Let $T = \{\mathrm{diag}(u_1(a), \ldots, u_m(a)) \mid a \in \mathscr{D}_n^*(K)\} \subseteq \mathscr{D}_m^*(K)$. Then $0 \in \overline{T}$ in $\mathscr{D}_n(K)$ and $\mathscr{X}(\overline{T}) \cong \langle u_1, \ldots, u_n \rangle$. Grillet [26] has shown that any finitely generated, totally cancellative, commutative semigroup can be embedded in a free commutative semigroup. Thus we have the following result (see [66] for further details).

<u>Theorem 8.14</u>. There is a contravariant equivalence between the category of connected diagonal monoids with zero and the category of finitely generated, totally cancellative, commutative semigroups without idempotents.

9 CROSS–SECTION LATTICES

In this chapter we introduce the central notion of cross–section lattices, due to the author [72], [74], [76].

Definition 9.1. Let M be a connected monoid with group of units G. Then $\Lambda \subseteq E(M)$ is a weak cross–section lattice if

(i) $|\Lambda \cap J| = 1$ for all $J \in \mathcal{U}(M)$.

(ii) If $e,f \in \Lambda$, then $J_e \geq J_f$ implies $e \geq f$.

If further $\Lambda \subseteq E(\overline{T})$ for some maximal torus T of G, then Λ is a cross–section lattice (which is necessarily a sublattice of $E(\overline{T})$).

Example 9.2. In $\mathcal{M}_2(K)$, $\Lambda = \left\{\begin{bmatrix}1 & 0\\0 & 1\end{bmatrix}, \begin{bmatrix}1 & 0\\0 & 0\end{bmatrix}, \begin{bmatrix}0 & 1\\0 & 1\end{bmatrix}, \begin{bmatrix}0 & 0\\0 & 0\end{bmatrix}\right\}$ is a weak cross–section lattice which is not a cross–section lattice.

The following result is due to the author [72].

Theorem 9.3. Let M be a connected monoid with zero and group of units G. Let T be a maximal torus of G. Then $\Lambda = \{e \in E(\overline{T}) \mid \text{for all } f \in E(M),\ e \mathcal{R} f \text{ implies } f \in \overline{B}\}$ is a cross–section lattice of M for any Borel subgroup B of G containing T.

Proof. Let $e,f \in \Lambda$ such that $e \mathcal{J} f$ in M. Then by Theorem 5.9, there exist $e',f' \in E(M)$ such that $e \mathcal{R} e' \mathcal{L} f' \mathcal{R} f$. Since $e,f \in \Lambda$, $e',f' \in \overline{B}$. Thus $e \mathcal{J} f$ in $\overline{B}$. Since B is solvable, we see by Corollary 6.32 that $e = f$. Hence $|J \cap \Lambda| \leq 1$ for all $J \in \mathcal{U}(M)$. For $e \in E(M)$, let $X_e = \{f \in E(M) | e \mathcal{R} f\}$. Now let $J_1,J_2 \in \mathcal{U}(M)$, $J_1 \geq J_2$. By Proposition 6.25, there exist $e_i \in J_i \cap E(T)$ such that $e_1 \geq e_2$. Extend $\{e_1,e_2\}$ to a maximal chain Γ of $E(T)$. By Proposition 6.24, $C_G^r(\Gamma)$ is a connected solvable subgroup of G. Thus there exists a Borel subgroup B' of G such that $C_G^r(\Gamma) \subseteq B'$. By Corollary 6.18, $X_{e_1}, X_{e_2} \subseteq \overline{B'}$. Now $T \subseteq B \cap B'$. By Theorem 4.22, there exists $u \in N_G(T)$ such that $u^{-1}B'u = B$. Let $f_i = u^{-1}e_iu$, $i = 1,2$. Then $f_1,f_2 \in E(T)$, $f_1 \geq f_2$. Also $X_{f_1}, X_{f_2} \subseteq \overline{B}$. Thus $f_1,f_2 \in \Lambda$. Since $f_i \in J_i$, $i = 1,2$, the proof is complete.

Corollary 9.4. Let M be a connected monoid with group of units G and let T be a maximal torus of G. Then for any chain Γ in $E(T)$, there exists a cross–section lattice Λ of M such that $\Gamma \subseteq \Lambda \subseteq E(T)$.

Proof. Let η denote the zero of $E(T)$. Then $E(T) = E(T_\eta)$. By Theorem 6.16, T_η is a maximal torus of G_η. By Proposition 6.27, any cross–section lattice of M_η is a cross–section lattice of M. Hence we may assume that $\eta = 0$ is the zero of M. Now $\Gamma \subseteq \Gamma'$ for some maximal chain Γ' of $E(T)$. By Proposition 6.24, $C_G^r(\Gamma') \subseteq B$ for some Borel subgroup B of G. By Corollary 6.18 and Theorem 9.3, $\Gamma' \subseteq \Lambda$ for some cross–section lattice $\Lambda \subseteq E(T)$.

The following result is due to the author [70].

Theorem 9.5. Let M be a connected monoid with zero 0 and group of units G. Then G is solvable if and only if $\mathcal{U}(M)$ is relatively complemented.

Proof. Let T be a maximal torus of G. If G is solvable, then by Theorem 6.20, Proposition 6.25, Corollary 6.32, $\mathcal{U}(M) \cong E(\overline{T})$ is relatively complemented. So assume conversely that $\mathcal{U}(M)$ is relatively complemented. We prove by induction on dim M that G is solvable. Let $e \in E(\overline{T})$, $e \neq 0,1$. By Proposition 6.27, $\mathcal{U}(M_e)$, $\mathcal{U}(eMe)$ are relatively complemented. So by the induction hypothesis, M_e, eMe have solvable groups of units. Thus by Proposition 6.27, Corollary 6.32, we have,

$$\begin{gathered}\text{if } e,f_1,f_2 \in E(\overline{T}),\ e \neq 0,1,\ f_1 \mathcal{J} f_2 \text{ and if} \\ \text{either } e \geq f_i,\ i = 1,2 \text{ or if } e \leq f_i,\ i = 1,2, \text{ then } f_1 = f_2\end{gathered} \tag{9}$$

By Corollary 6.32, it suffices to show that $w(J) = 1$ for all $J \in \mathcal{U}(M)$. Suppose not. Then choose a maximal $J_o \in \mathcal{U}(M)$ with $w(J_o) > 1$. Let $e_o, e'_o \in J_o \cap E(\overline{T})$, $e_o \neq e'_o$. Let $J' \in \mathcal{U}(M)$ such that J' covers J_o. Then $w(J) = 1$. Let $J' \cap E(\overline{T}) = \{\eta\}$. By Proposition 6.25, $\eta \geq e_o$, $n \geq e'_o$. So by (9), $\eta = 1$. Thus $J' = G$ and G covers J_o. So by Corollary 6.22, 1 covers e_1, e_2. Let $\dim T = p$, $X = \{e \in E(\overline{T}) \mid ht(e) = p-1\}$, $Y = \{f \in E(\overline{T}) \mid ht(f) = p-2\}$. Then e_o, $e'_o \in X$. So by Corollary 8.10, there exist $e_1,\ldots,e_{k+1} \in X$, $f_1,\ldots,f_k \in Y$ such that $e_o = e_1$, $e'_o = e_{k+1}$, $e_i > f_i$, $e_{i+1} > f_i$, $i = 1,\ldots,k$. By Corollary 9.4, there exists a cross-section lattice Λ of M such that $e_o = e_1 \in \Lambda \subseteq E(\overline{T})$. Suppose $p > 2$. Then $Y \neq \{0\}$. We will obtain a contradiction. There exists $f_1' \in \Lambda$ such that $f_1 \mathcal{J} f_1'$, $e_1 > f_1'$. By (9), $f_1 = f_1' \in \Lambda$. So there exists $e'_2 \in \Lambda$ such that $e_2 \mathcal{J} e'_2$, $e'_2 > f_1$. So again by (9), $e_2 = e'_2 \in \Lambda$. Continuing, we find that $e'_o \in \Lambda$. Since $e_o \in \Lambda$, $e_o \mathcal{J} e'_o$, $e_o \neq e'_o$, we have a contradiction. Hence $p = 2$. Then by Remark 8.8, $E(\overline{T}) = \{1, e_o, e'_o, 0\}$. Hence $\mathcal{U}(M) = \{G,J,0\}$ is not relatively complemented. This proves the theorem.

The following result is from the author [74].

Proposition 9.6. Let M be a connected regular monoid and let Λ_1, Λ_2 be two weak cross–section lattices of M. If $\Lambda_1 \cap \Lambda_2$ contains a maximal chain of $E(M)$, then $\Lambda_1 = \Lambda_2$.

Proof. Let G denote the group of units of M and let $ht(M) = 0$. Let $\phi_i: \mathcal{U}(M) \to \Lambda_i$ be the bijections given by $\phi_i(J) \in J$, $i = 1,2$. We prove by induction on $\dim M$ that $\phi_1 = \phi_2$. Let $\Gamma = \{1 > e > ...\}$ be a maximal chain of $E(M)$ contained in $\Lambda_1 \cap \Lambda_2$. Let J denote the $\mathcal{J}$–class of e. If $f \in \Lambda_i$, then let $\Lambda_i(f) = \{h \in \Lambda_i \mid f \geq h\}$. Then by Proposition 6.27, $\Lambda_i(f)$ is a weak cross–section lattice of fMf. Now eMe is regular and $\Gamma' = \Gamma\backslash\{1\}$ is a maximal chain in $E(eMe)$ with $\Gamma' \subseteq \Lambda_1(e) \cap \Lambda_2(e)$. Hence $\Lambda_1(e) = \Lambda_2(e)$. Suppose there exists $h \in \Lambda_1\backslash\Lambda_2$. Then there exists $e' \in E(M)$, $ht(e') = p-1$, $e' \geq h$. Let J' denote the $\mathcal{J}$–class of e'. Then $J \neq J'$. By Corollary 8.11, there exist distinct $J = J_o, J_1,...,J_{t+1} = J' \in \mathcal{U}(M)$, distinct $J_o^*,...,J_t^* \in \mathcal{U}(M)$ such that $J_k > J_k^*$, $J_{k+1} > J_k^*$, $ht(J_k) = p-1$, $ht(J_k^*) = p-2$, $k = 0,...,t$. Then $\phi_i(J_k) > \phi_i(J_k^*)$, $\phi_i(J_{k+1}) > \phi_i(J_k^*)$, $i = 1,2$, $k = 0,...,t$. Now $\phi_1(J_o) = \phi_2(J_o) = e$. Let $f = \phi_1(J_o^*)$. Then $f \in \Lambda_1(e) = \Lambda_2(e)$. So $f = \phi_2(J_o^*)$. Extend $\{1 > e > f\}$ to a maximal chain Γ_1 of $\Lambda_1(e) = \Lambda_2(e)$. Let $e_1 = \phi_1(J_1)$, $e_1' = \phi_2(J_1)$. Then $e_1 > f$, $e_1' > f$. Since $J_o \neq J_1$, we see by Theorem 9.5 that G_f is solvable. By Theorem 7.4, G_f is reductive. Hence G_f is a torus. So by Remark 8.8, $|E(M_f)| = 4$. Since $1,e,e_1,e_1',f \in M_f$, $e_1 = e_1'$. Let $\Gamma_2 = (\Gamma_1\backslash\{e\}) \cup \{e_1\}$. Then Γ_2 is a maximal chain of $E(e_1Me_1)$ and $\Gamma_2 \subseteq \Lambda_1(e_1) \cap \Lambda_2(e_1)$. By the induction hypothesis $\Lambda_1(e_1) = \Lambda_2(e_1)$. Continuing this process, we see that $e' \in \Lambda_2$ and $\Lambda_1(e') = \Lambda_2(e')$. Hence $h \in \Lambda_1(e') \subseteq \Lambda_2$. This contradiction completes the proof.

Corollary 9.7. Let M be a connected regular monoid. Then every weak cross–section lattice of M is a cross–section lattice and any two cross–section lattices are conjugate.

Proof. Let G denote the group of units of M and let Λ be a weak cross-section lattice of M. Let Γ be a maximal chain of Λ. Then Γ is a maximal chain of $E(M)$. By Corollary 6.10, $\Gamma \subseteq \overline{T}$ for some maximal torus T of G. So by Corollary 9.4, $\Gamma \subseteq \Lambda' \subseteq \overline{T}$ for some cross-section lattice Λ of M. By Proposition 9.6, $\Lambda = \Lambda'$. Now let Λ_1 be a cross-section lattice of M. There exists a maximal chain Γ_1 of Λ_1 such that for all $e \in \Gamma$, there exists $f \in \Gamma_1$ such that $e \mathcal{J} f$. By Corollary 6.8, Theorem 6.16 (ii), there exists $x \in \dot{G}$ such that $x^{-1}\Gamma_1 x = \Gamma$. So $\Gamma \subseteq \Lambda \cap x^{-1}\Lambda_1 x$. Hence $\Lambda = x^{-1}\Lambda_1 x$ by Proposition 9.6.

Proposition 9.8. Let M be a connected regular monoid with zero 0 and group of units G. Let B be a Borel subgroup of G, $e \in E(\overline{B})$. Then the following conditions are equivalent.

(i) $B \subseteq C_G^r(e)$.

(ii) For any $f \in E(M)$, $e \mathcal{R} f$ implies $f \in E(\overline{B})$.

(iii) For any $f \in E(\overline{B})$, $e \mathcal{J} f$ in $\overline{B}$ implies $e \mathcal{R} f$.

Proof. For $f \in E(M)$, let $X_f = \{h \in E(M) \mid f \mathcal{R} h\}$. Let T be a maximal torus of B with $e \in \overline{T}$. Let Γ be a maximal chain in $E(\overline{T})$ with $e \in \Gamma$. Then $B_1 = \overline{C_G^r(\Gamma)}$ is a Borel subgroup of G by Theorem 7.1. Clearly $B_1 \subseteq C_G^r(e)$.

(i) ⇒ (ii). B, B_1 are Borel subgroups of $C_G^r(e)$ and the width of e in $C_G^r(e)$ is 1. So by Lemma 6.28, there exists $u \in C_G(e)$ such that $u^{-1}B_1u = B$. Then $X_e = u^{-1}X_eu \subseteq \overline{B}$.

(i) ⇒ (iii). Let $f \in E(\overline{B})$, $e \mathcal{J} f$ in B. By Corollary 6.8, $xex^{-1} = f$ for some $x \in B \subseteq C_G^r(e)$. So $xe = exe$ and $ef = f$. Thus $e \mathcal{R} f$ by Theorem 1.4.

(iii) ⇒ (i). Let $x \in B$. Then $xex^{-1} \in E(\overline{B})$, $e \mathcal{J} xex^{-1}$ in $\overline{B}$. So $e \mathcal{R} xex^{-1}$. Hence $exex^{-1} = xex^{-1}$ and $xe = exe$. Thus $B \subseteq C_G^r(e)$.

(ii) ⇒ (i). Since $T \subseteq B \cap B_1$, we see by Theorem 4.22 that $x^{-1}B_1x = B$ for some $x \in N_G(T)$. Let $f = x^{-1}ex \in E(\overline{T})$. By assumption $X_e \subseteq \overline{B}$. By Corollary

6.18, $X_e \subseteq \overline{B}_1$. Hence $X_f = u^{-1}X_e u \subseteq \overline{B}$. By Theorem 5.9, there exists $e_1, f_1 \in E(M)$ such that $e \mathcal{R} e_1 \mathcal{L} f_1 \mathcal{R} f$. So $e_1 \in X_e \subseteq \overline{B}$, $f_1 \in X_f \subseteq \overline{B}$. Hence $e \mathcal{J} f$ in $\overline{B}$. Since B is solvable, we see by Corollary 6.32 that $e = f$. Hence $x \in C_G(e)$ and $B = x^{-1}B_1 x \subseteq C_G^r(e)$. This completes the proof.

Let M be a connected regular monoid with zero 0 and group of units G. Then G is a reductive group. Let T be a maximal torus of G. As in Definition 4.21, let $\mathcal{B}(T)$ denote the set of all Borel subgroups of G containing T. If $B \in \mathcal{B}(T)$, then let $B^- \in \mathcal{B}(T)$ denote the opposite Borel subgroup of G relative to T, i.e. $B \cap B^- = T$. Let

$$\mathcal{C}(T) = \{\Lambda \mid \Lambda \subseteq E(\overline{T}) \text{ is a cross-section lattice of } M\}.$$

<u>Definition 9.9</u>. If $B \in \mathcal{B}(T)$, then the <u>cross-section lattice</u> of B, $\xi(B) = \{e \in E(\overline{T}) \mid ae = eae \text{ for all } a \in B\}$ and the <u>opposite cross-section lattice of</u> B, $\xi^-(B) = \{e \in E(\overline{T}) \mid ea = eae \text{ for all } a \in B\}$. If $\Lambda \in \mathcal{C}(T)$, then the <u>Borel subgroup of</u> Λ, $\beta(\Lambda) = C_G^r(\Lambda)$ and the <u>opposite Borel subgroup</u> of Λ, $\beta^-(\Lambda) = C_G^{\ell}(\Lambda)$.

The fundamental theorem of cross-section lattices due to the author [72], [74], [76] is:

<u>Theorem 9.10</u>. Let M be a connected regular monoid with zero and group of units G. Let T be a maximal torus of G. Let $\mathcal{B} = \mathcal{B}(T)$, $\mathcal{C} = \mathcal{C}(T)$. Then

(i) If $\Lambda \in \mathcal{C}$, then $\beta^-(\Lambda) \in \mathcal{B}$ and $\beta(\Lambda)^- = \beta^-(\Lambda)$.

(ii) If $B \in \mathcal{B}$, then $\xi(B), \xi^-(B) \in \mathcal{C}$ and $\xi(B^-) = \xi^-(B)$.

(iii) $\beta = \xi^{-1}$ and $\beta^- = (\xi^-)^{-1}$.

(iv) If $\Lambda_1, \Lambda_2 \in \mathcal{C}$, then $\Lambda_1^{\sigma} = \Lambda_2$ for some $\sigma \in W$.

<u>Proof</u>. Let $B \in \mathcal{B}$ and let $\Lambda = \xi(B)$. By Theorem 9.3 and Proposition 9.8, $\Lambda \in \mathcal{C}$. Let Γ be a maximal chain in Λ. Then Γ is a maximal chain in E(M) and hence

by Theorem 7.1, $C_G^r(\Gamma) \in \mathscr{B}$. Clearly $B \subseteq \beta(\Lambda) = C_G^r(\Lambda) \subseteq C_G^r(\Gamma) \in \mathscr{B}$. So $B = \beta(\Lambda) = C_G^r(\Gamma)$. Hence $B = \beta(\xi(B))$. By Theorem 7.1, $B^- = C_G^\ell(\Gamma)$. Let $\Lambda' = \xi^-(B^-)$. Then as above, $\Lambda' \in \mathscr{C}$, $\beta^-(\Lambda') = B^-$. Clearly $\Gamma \subseteq \Lambda'$. By Proposition 9.6, $\Lambda = \Lambda'$. Thus $\beta^-(\Lambda) = \beta(\Lambda^-)$ and $\xi^-(B^-) = \xi(B)$. Next let $\Lambda \in \mathscr{C}$, $\Gamma \subseteq \Lambda$ a maximal chain. Let $B = C_G^r(\Gamma) \in \mathscr{B}$ Clearly $\Gamma \subseteq \xi(B)$. So by Proposition 9.6, $\xi(B) = \Lambda$. By the above, $\beta(\Lambda) = \beta(\xi(B)) = B$. Similarly $\beta^-(\Lambda) \in \mathscr{B}$ and $\xi^-(\beta^-(\Lambda)) = \Lambda$. Finally let $\Lambda_1, \Lambda_2 \in \mathscr{C}$. Then $\beta(\Lambda_1), \beta(\Lambda_2) \in \mathscr{B}$ By Theorem 4.22, there exists $\sigma \in W$ such that $\beta(\Lambda_1^\sigma) = \beta(\Lambda_1)^\sigma = \beta(\Lambda_2)$. So $\Lambda_1^\sigma = \Lambda_2$. This proves the theorem.

The following result of Renner [97] will be needed in Chapter 11.

Corollary 9.11. Let M be a connected regular monoid with zero 0, group of units G. Let T be a maximal torus of G. Suppose $|W(G)| = 2$, $\mathscr{B}(T) = \{B, B^-\}$, $U = B_u$, $U^- = B_u^-$. Then for any $e \in E(\overline{T})$ with $e \notin C(M)$, either $e\,U = U^-e = \{e\}$ or else $U\,e = e\,U^- = \{e\}$.

Proof. Let $\Lambda = \xi(B)$, $\Lambda^- = \xi(B^-)$. By Theorem 9.10, $\mathscr{C}(T) = \{\Lambda, \Lambda^-\}$. By Corollary 9.4, $E(\overline{T}) = \Lambda \cup \Lambda^-$. Suppose $e \in \Lambda$. Then $U\,e = e\,U\,e$, $e\,U^- = e\,U^-e$. Since $e \notin C(M)$, $|W(C_G(e))| = 1$ by Proposition 6.25. Thus $C_G(e)$ is a torus. So by Theorem 6.16, the $\mathscr{H}$-class H of e is a torus. We have a homomorphism $\phi\colon U \to H$ given by $\phi(x) = xe = exe$. Since U is unipotent, we see that $Ue = \{e\}$. Similarly $e\,U^- = \{e\}$. In the same way, $e \in \Lambda^-$ implies $e\,U = U^-e = \{e\}$.

As another application of cross-section lattices, we prove the following result of the author [77].

Corollary 9.12. Let M be a connected regular monoid with zero 0 and group of units G. Let $e, e' \in E(M)$ with $ht(e) = ht(e') = p > 0$. Then there exist $e = e_o$, $e_1, \ldots, e_k = e'$, $f_1, \ldots, f_k \in E(M)$ such that $ht(e_i) = p$, $ht(f_i) = p - 1$, $e_i > f_i$, $e_{i-1} > f_i$,

$i = 1,\ldots,k$.

Proof. By Corollary 8.10, the theorem is true when G is a torus. Thus by Corollary 6.10, we are reduced to the case when $e \mathcal{J} e'$. By Theorem 5.9, we are then further reduced to the case when $e \mathcal{R} e'$ or $e \mathcal{L} e'$. By symmetry assume $e \mathcal{R} e'$. Let Λ be a cross–section lattice of M with $e \in \Lambda$. Let $B = C_G^r(\Lambda)$, $T = C_G(\Lambda)$. Then B is a Borel subgroup of G, T a maximal torus of G, $T \subseteq B$, $e \in E(T)$, $e' \in \overline{B}$. By Theorem 9.10, there exists a cross–section lattice $\Lambda' \subseteq E(T)$ such that $B = C_G^\ell(\Lambda')$. There exists $u \in \Lambda'$ such that $e \mathcal{J} u$ in M. Then $ht(e) = ht(u)$. So there exist $e = u_0, u_1,\ldots,u_k = u, v_1,\ldots,v_k \in E(T)$ such that $ht(u_i) = p$, $ht(v_i) = p - 1$, $u_i > v_i$, $u_{i-1} > v_i$, $i = 1,\ldots,k$. By Corollary 6.18, $e, e', v_1 \in \overline{C_B^r(e,v_1)}$. There exists $x \in C_B^r(e,v_1)$ such that $e' = x^{-1}ex$. Let $v_1' = x^{-1}v_1x \in E(\overline{B})$. Then $v_1 \mathcal{R} v_1'$, $e' \geq v_1'$. Now $v_1, v_1', u_2 \in \overline{C_B^r(u_1,v_1)}$. There exists $y \in C_B^r(u_1,v_1)$ such that $v_1' = y^{-1}v_1y$. Let $u_1' = y^{-1}u_1y \in E(\overline{B})$. Then $u_1' \geq v_1'$, $u_1 \mathcal{R} u_1'$. Continuing, we find $e' = u_0', u_1',\ldots,u_k' = u', v_1',\ldots,v_k' \in E(\overline{B})$ such that $ht(u_i') = p$, $ht(v_i') = p - 1$, $u_i' > v_i'$, $u_{i-1}' > v_i'$, $u_i \mathcal{R} u_i'$, $v_i \mathcal{R} v_i'$, $i = 1,\ldots,k$. In particular, $u \mathcal{R} u'$ and $u' = zuz^{-1}$ for some $z \in C_B^r(u)$. Since $B \subseteq C_G^\ell(u)$, we see that $u = u'$. This proves the theorem.

Problem 9.13. Is Corollary 9.12 true without the assumption of regularity?

10 $\mathscr{E}$–STRUCTURE

Let M be a connected regular monoid with zero 0 and group of units G. Fix a maximal torus T of G. As usual $W = N_G(T)/T$ is the Weyl group.

Definition 10.1. (i) $\mathscr{E} = \mathscr{E}(M) = (E(\overline{T}), \leq, \sim)$ where for $e,f \in E$, $e \sim f$ if $e \mathscr{J} f$. Let ht $\mathscr{E}$ = dim T = the length of any maximal chain in $\mathscr{E}$.

(ii) A bijection $\sigma: e \to e^{\sigma}$ of $\mathscr{E}$ is an automorphism if $e \sim e^{\sigma}$ for all $e \in \mathscr{E}$, and for all $e,f \in \mathscr{E}$, $e \geq f$ if and only if $e^{\sigma} \geq f^{\sigma}$. Let $\mathscr{W} = \mathscr{W}(\mathscr{E})$ denote the group of all automorphisms of $\mathscr{E}$.

(iii) $\mathscr{U}' = \mathscr{U}'(M) = \mathscr{E}/\sim$. If $A_1, A_2 \in \mathscr{U}'$, then define $A_1 \leq A_2$ if there exist $e_i \in A_i$, $i = 1,2$ such that $e_1 \leq e_2$.

By Proposition 6.25 we have,

Proposition 10.2. $\mathscr{U}'$ is a finite lattice isomorphic to $\cdot\, \mathscr{U}$.

We refer to $\mathscr{E}$ as the $\mathscr{E}$–structure of M, $\mathscr{U}' \cong \mathscr{U}$ as the $\mathscr{U}$–structure of M. These are finite combinational structures, with $\mathscr{E}$ determining $\mathscr{U}$.

Example 10.3. Let $G_0 = \{A \otimes (A^{-1})^t \mid A \in SL(3,K)\}$, $M = \overline{KG_0} \subseteq \mathscr{M}_9(K)$. Let $e_1 = \text{diag}(1,0,0)$, $e_2 = \text{diag}(0,1,0)$, $e_3 = \text{diag}(0,0,1)$, $f_1 = e_2 + e_3$, $f_2 = e_1 + e_3$, $f_3 = e_1 + e_2$. Then by Example 8.5, $\mathscr{E}(M)$, or more precisely $\mathscr{U}'$, is represented by the following diagram:

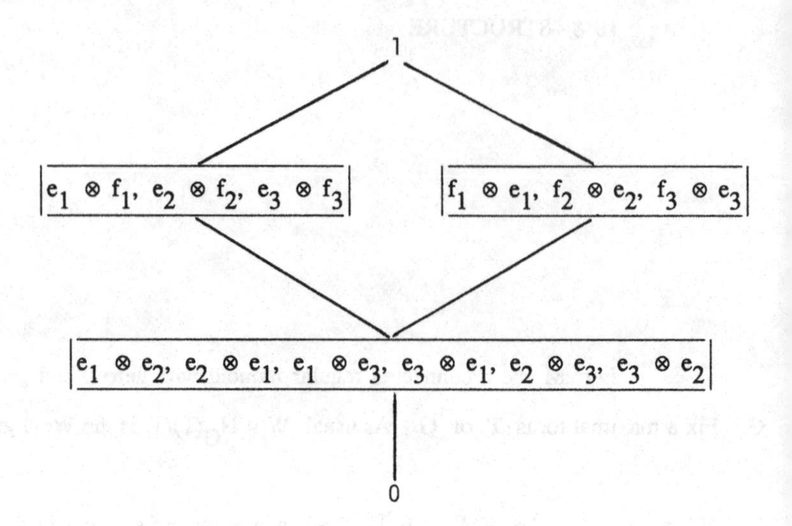

Example 10.4. Let $G_o = \{A \oplus (A^{-1})^t | A \in SL(3,K)\}$, $M = \overline{KG_o} \subseteq \mathcal{M}_6(K)$. Let $e_1 = \text{diag}(1,0,0)$, $e_2 = \text{diag}(0,1,0)$, $e_3 = \text{diag}(0,0,1)$. Then by Example 8.6, $\mathcal{E}(M)$ is represented by the following diagram:

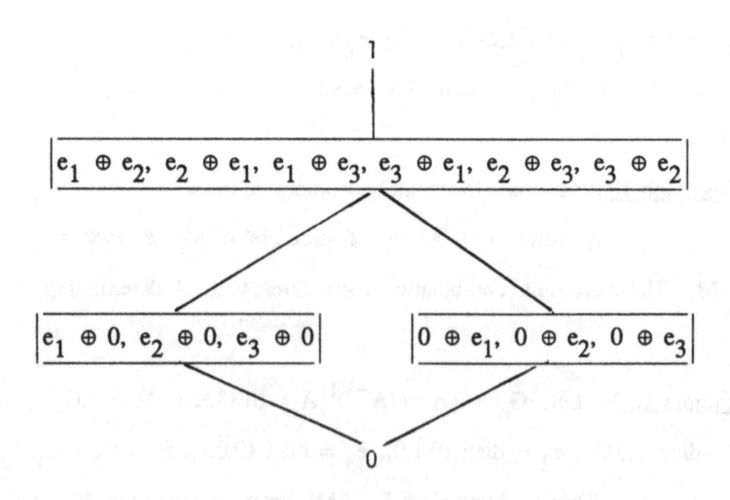

We refer to [72] for further diagrams of $\mathcal{E}$–structures.

Definition 10.5. If $e \in \mathscr{E}$, then let $\mathscr{E}_e = \{f \in \mathscr{E} | f \geq e\}$, $e\,\mathscr{E} = \{f \in \mathscr{E} | f \leq e\}$. We consider $\mathscr{E}_e$, $e\,\mathscr{E}$ with $\leq$, $\sim$ restricted to them.

If $e \in \mathscr{E}$, H the $\mathscr{H}$–class of e, then by Theorem 6.16, $e\,T$, T_e are maximal tori of H, G_e, respectively. By Proposition 6.27, we have,

Proposition 10.6. If $e \in \mathscr{E}$, then $\mathscr{E}(M_e) = \mathscr{E}_e$, $\mathscr{E}(eMe) = e\,\mathscr{E}$.

The following result is due to the author [72].

Theorem 10.7. $\mathscr{W} \cong W$.

Proof. W acts on $\mathscr{E}$ faithfully by Theorem 7.1. In this way $W \subseteq \mathscr{W}$. We prove by induction on $\dim M$ that $W = \mathscr{W}$. Let $e \in \mathscr{E}$ such that 1 covers e. Let $X = \{f | f \in \mathscr{E}, e \sim f\}$. So $|X| = w(e)$. By Proposition 6.25, W and hence $\mathscr{W}$ acts transitively on X. By Theorem 6.16, Proposition 6.25,

$$\begin{aligned} |W| &= w(e)|W(eMe)| \cdot |W(M_e)| \\ &= w(e)|W(eMe)| \end{aligned}$$

Let $\mathscr{W}_o = \{\sigma \in \mathscr{W} | e^\sigma = e\}$. Then

$$|\mathscr{W}| = w(e)\ |\mathscr{W}_o|$$

If $\sigma \in \mathscr{W}_o$, let $\sigma \in \mathscr{W}(e\,\mathscr{E})$ denote the restriction of σ to $e\mathscr{E}$. Clearly the map $\sigma \to \bar{\sigma}$ is a homomorphism. If $\bar{\sigma} = 1$, then σ fixes a maximal chain in $\mathscr{E}$ and hence by Corollary 8.13, $\sigma = 1$. Thus $|\mathscr{W}_o| \leq |\mathscr{W}(e\mathscr{E})|$. By the induction hypothesis, $|\mathscr{W}(e\mathscr{E})| = |W(eMe)|$. Hence $|\mathscr{W}| \leq |W|$. Thus $W = \mathscr{W}$, proving the theorem.

From now on we identify W with $\mathscr{W}$.

Definition 10.8. (i) If $\Gamma \subseteq \mathscr{E}$, $V \subseteq W$, then $C_V(\Gamma) = \{\sigma \in V \mid e^\sigma = e$ for all $e \in \Gamma\}$.

(ii) If $e \in \mathscr{E}$, $\sigma \in C_W(e)$, then let $e\sigma \in W(eMe)$, $\sigma_e \in W(M_e)$ denote the restrictions of σ to $e\mathscr{E}$, $\mathscr{E}_e$, respectively.

Proposition 10.9. Let $e \in \mathscr{E}$. Then

(i) $C_W(e) = W(C_G(e))$

(ii) $W(eMe) = \{e\sigma \mid \sigma \in C_W(e)\}$

(iii) $W(M_e) = \{\sigma_e \mid \sigma \in C_W(e)\}$

(iv) If $\sigma \in C_W(e)$ with $e\sigma = \sigma_e = 1$, then $\sigma = 1$.

(v) $C_W(e) \cong W(eMe) \times W(M_e)$.

Proof. Since $T \subseteq C_G(e)$, (i) is clear. Define $\phi: C_W(e) \to W(eMe)$ as $\phi(\sigma) = e\sigma$, $\psi: C_W(e) \to W(M_e)$ as $\psi(\sigma) = \sigma_e$. Then σ, ψ are homomorphisms. If $e\sigma = \sigma_e = 1$, then σ fixes a maximal chain in $\mathscr{E}$ and hence by Theorem 7.1, $\sigma = 1$. By Theorem 6.16, $|C_W(e)| = |W(eMe)| \cdot |W(M_e)|$. The result follows.

The following result is due to the author [72].

Theorem 10.10. Let $\sigma \in W$, $\sigma \neq 1$. Then σ is a reflection if and only if σ fixes a chain Γ in $\mathscr{E}$ of length ht $\mathscr{E}$–1. In such a case $C_G(T_\sigma) = C_G(\Gamma)$.

Proof. Let $m = \dim T = \text{ht } \mathscr{E}$. First suppose that σ fixes a chain Γ in $\mathscr{E}$ of length $m-1$. Let $T_\sigma = \{t \in T \mid t^\sigma = t\}^c$. Then by Lemma 6.14, $\Gamma \subseteq \overline{T}_\sigma$. So by Theorem 6.20, $m - 1 \leq \dim T_\sigma < \dim T$. So $\dim T_\sigma = m - 1$. Since $\sigma \in W(C_G(T_\sigma))$, we see by Theorem 4.45 that σ is a reflection. Let $T_o = \text{rad } C_G(\Gamma)$. Since $\sigma \in W(C_G(\Gamma))$, $T_o \subseteq T_\sigma$. But $\Gamma \subseteq \overline{T}_o$ by Corollary 6.31. So $T_o = T_\sigma$ by Theorem 6.20. Thus $C_G(\Gamma) = C_G(T_\sigma)$.

Conversely assume that σ is a reflection. By Theorem 4.45, $\dim T_\sigma = m - 1$. By Lemma 6.14, $0 \in \overline{T}_\sigma$. By Theorem 6.20, there exists a chain Γ

of $E(T_\sigma)$ of length $m-1$. This proves the theorem.

Let $e \in \mathscr{E}$, $\sigma \in W$, a reflection such that $e^\sigma = e$. Then by Lemma 6.14, $e \in T_\sigma$. By Theorem 6.20, there exists a chain $\Gamma \subseteq T_\sigma$ with $e \in \Gamma$ such that Γ has length $\dim T - 1$. Hence σ fixes a maximal chain in either $E(T_e)$ or $E(eT)$. Thus by Corollary 8.13, either $\sigma_e = 1$ or $e\sigma = 1$. Thus we have shown,

Corollary 10.11. Let $\sigma \in W$ be a reflection, $e \in \mathscr{E}$, $e^\sigma = e$. Then either $e\sigma = 1$, σ_e is a reflection in $W(M_e)$ or else $\sigma_e = 1$ and $e\sigma$ is a reflection in $W(eMe)$.

The following result is due to Renner [91].

Theorem 10.12. Let M, M' be connected monoids with zero, M regular, $\phi: M \to M'$ a homomorphism such that $M' = \overline{\phi(M)}$ and $\phi^{-1}(0) = \{0\}$. Then

(i) $M' = \phi(M)$ is regular.

(ii) ϕ is idempotent separating.

(iii) $\phi: \mathscr{E}(M) \cong \mathscr{E}(M')$.

Proof. Let G, G' be the groups of units of M, M', respectively. By Theorem 2.21, $\dim M = \dim M'$. So $\phi(G) = G'$. By Theorem 7.3, M' is regular. Let T be a maximal torus of G. First we show that ϕ is 1–1 on $E(\overline{T})$. So let $e,f \in E(\overline{T})$, $\phi(e) = \phi(f)$, $e \neq f$. Then $\phi(ef) = \phi(f) \neq 0$. Suppose $e > ef$. Then since $E(\overline{T})$ is relatively complemented, there exists $h \in E(\overline{T})$ such that $e > h > 0$, $efh = 0$. Then $0 = \phi(efh) = \phi(eh) = \phi(h)$, a contradiction. Thus $e = ef$. Similarly $f = ef$. Hence $e = f$ and ϕ is 1–1 on $E(\overline{T})$. Now let $e,f \in E(\overline{T})$ such that $\phi(e) \geq \phi(f)$. Then $\phi(ef) = \phi(f)$ and hence $ef = f$. Thus $E(\overline{T}) \cong \phi(E(\overline{T})) \subseteq E(\overline{\phi(T)})$. By Corollary 8.12, $E(\overline{T}) = E(\overline{\phi(T)})$. So $M' = G'E(\overline{\phi(T)})G' = \phi(M)$. Now let $e,f \in E(M)$, $\phi(e) = \phi(f)$. By Lemma 7.6, there exists a maximal torus T_1 of G, $e_1, f_1 \in E(\overline{T_1})$ such that $e \mathscr{R} e_1$, $f \mathscr{R} f_1$. Then $\phi(e_1) \mathscr{R} \phi(f_1)$ and hence $\phi(e_1) = \phi(f_1)$. So $e_1 = f_1$. Thus

$e \mathcal{R} f$. Similarly $e \mathcal{L} f$. Hence $e = f$. Thus ϕ is idempotent separating. Finally let $e,f \in E(T)$ such that $\phi(e) \mathcal{J} \phi(f)$. Then by Proposition 1.19, $e \mathcal{J} f$. It follows that $\mathcal{E}(M) \cong \mathcal{E}(M')$.

The situation in Theorem 10.12 arises quite often. For example, we have the following from [76; Proposition 2.2].

Corollary 10.13. Let M be a connected regular monoid with zero 0 and group of units G such that $\text{rad } G$ is one dimensional. Then there exists an idempotent separating homomorphism $\phi: M \to \mathcal{M}_p(K)$ such that $\text{rad } \phi(G)$ consists of scalar matrices.

Proof. We may assume that M is a closed submonoid of $\text{End}(V)$, for some finite dimensional vector space V over K. We can further assume that M contains the zero of $\text{End}(V)$. Let $T_o = \text{rad } G$. Then $\dim T_o = 1$, $0 \in \overline{T}_o$. Let $\chi_1,\dots,\chi_m \in \mathcal{X}(T_o)$ denote the weights of T_o. Let $V_i = \{v \in V \mid tv = \chi_i(t)v$ for all $t \in T_o\} \neq \{0\}$, $i = 1,\dots,m$. Then $V = V_1 \oplus \dots \oplus V_m$. Since T_o lies in the center of M, $MV_i \subseteq V_i$, $i = 1,\dots,m$. Now $\mathcal{X}(T_o)$ is a cyclic group. Since $0 \in \overline{T}_o$, there exist $n_1,\dots,n_m \in \mathbb{Z}^+$ such that $\chi_i^{n_i} = \chi_j^{n_j}$, $i,j = 1,\dots,m$. Let $V_i' = V_i \otimes \dots \otimes V_i$ denote the n_i^{th}-fold tensor product of V_i. If $a \in M$, then a acts on V_i' as

$$a(v_1 \otimes v_2 \otimes \dots) = av_1 \otimes av_2 \otimes \dots$$

Thus a acts on $V' = V_1' \oplus \dots \oplus V_m'$. Let $\phi(a)$ denote the corresponding element of $\text{End}(V')$. We therefore have a homomorphism $\phi: M \to \text{End}(V')$. Clearly $\phi^{-1}(0) = \{0\}$. By Theorem 10.12, ϕ is idempotent separating. If $t \in T_o$, then clearly $\phi(t) = \chi_i^{n_i}(t) \cdot 1$ is a scalar.

Let M be a connected regular monoid with zero 0, group of units G, T a maximal torus of G. Let $\mathcal{C} = \mathcal{C}(T)$, $\mathcal{B} = \mathcal{B}(T)$, $\beta, \beta^-, \xi, \xi^-$ be as in Definition 9.9. Note that if $\Lambda \subseteq \mathcal{E}$, then $\Lambda \in \mathcal{C}$ if and only if: (i) for all $e \in \mathcal{E}$, there exists a unique $f \in \Lambda$ such that $e \sim f$, (ii) for any $e,f \in \Lambda$, $\sigma \in W$, $e^\sigma \geq f$ implies $e \geq f$. Thus the family $\mathscr{C}$ is determined by $\mathscr{E}$. Now fix $\Lambda \in \mathscr{C}$ and let $B = \beta(\Lambda)$, $B^- = \beta^-(\Lambda) \in \mathscr{B}$

Definition 10.14. Let $\mathscr{S} = \mathscr{S}(\Lambda) = \mathscr{S}(B) = \{\sigma \mid 1 \neq \sigma \in W,\ B \cup B\sigma B \text{ is a group}\}$ denote the set of simple reflections relative to B. If $I \subseteq \mathscr{S}$, let $W_I = \langle I \rangle$, $P_I = BW_IB$, $\mathscr{P}(B) = \{P_I \mid I \subseteq \mathscr{S}\}$ the set of parabolic subgroups of G containing B. See Theorem 4.51.

We will show in Corollary 10.21 that $\mathscr{S}$ is determined within the system $\mathscr{E}$.

Lemma 10.15. Let $I \subseteq \mathscr{S}$, $P = BW_IB$, $\Gamma \subseteq \Lambda$. Then the following conditions are equivalent.

(i) $P = C^r_G(\Gamma)$

(ii) $I = C_{\mathscr{S}}(\Gamma)$

(iii) $W(P) = C_W(\Gamma)$

Proof. (i) $\Rightarrow$ (iii). Let $e \in \Gamma$. Then the width of e in $\bar{P}$ is 1. So $e^\sigma = e$ for all $\sigma \in W(P)$. So $W(P) \subseteq C_W(\Gamma)$. Now let $\sigma = xT \in C_W(\Gamma)$. Then $x \in C_G(\Gamma) \subseteq C^r_G(\Gamma) = P$. So $\sigma \in W(P)$

(iii) $\Rightarrow$ (ii). $I = \mathscr{S} \cap W_I = \mathscr{S} \cap C_W(\Gamma) = C_{\mathscr{S}}(\Gamma)$.

(ii) $\Rightarrow$ (i). Let $P_1 = C^r_G(\Gamma) \in \mathscr{P}$. By the above, $W(P_1) = C_W(\Gamma)$, $W(P_1) \cap \mathscr{S} = C_W(\Gamma) \cap \mathscr{S} = C_{\mathscr{S}}(\Gamma) = I$. So $P_1 = BW_IB = P$.

Lemma 10.16. Let $e \in \Lambda$, $e\Lambda = \{f \in \Lambda \mid e \geq f\}$, $\Lambda_e = \{f \in \Lambda \mid e \leq f\}$. Then $\Lambda_e \in \mathscr{C}(T_e)$ in M_e, $e\Lambda \in \mathscr{C}(eT)$ in eMe. If $\sigma \in C_{\mathscr{S}}(e)$, then either $\sigma_e = 1$ and $e\sigma \in \mathscr{S}(e\Lambda)$ or else $e\sigma = 1$ and $\sigma_e \in \mathscr{S}(\Lambda_e)$.

Proof. That $\Lambda_e \in \mathscr{C}(T_e)$, $e\Lambda \in \mathscr{C}(eT)$ follows from Proposition 6.27. By Corollary 7.2, B_e is a Borel subgroup of G_e and $Be = eBe$ is a Borel subgroup of the $\mathscr{H}$-class H of e. Clearly $Be \subseteq \beta(e\Lambda)$, $B_e \subseteq \beta(\Lambda_e)$. Thus $Be = \beta(e\Lambda)$, $B_e = \beta(\Lambda_e)$. Let $\sigma \in C_{\mathscr{S}}(e)$, $P = B \cup B\sigma B$. Then $W(P) = \{1,\sigma\}$, $Pe = ePe$. Since the width of e in $\overline{P}$ is 1, we see by Theorem 6.16 and Proposition 6.25 that $2 = |W(P)| = |W(P_e)| \cdot |W(Pe)|$. By Proposition 10.9, $W(P_e) = \{1,\sigma_e\}$, $W(Pe) = \{1,e\sigma\}$. Since $B_e \subseteq P_e$ and $Be \subseteq Pe$, we are done.

Lemma 10.17. Let $P \in \mathscr{P}(B)$ be a maximal parabolic subgroup, $T_o = T \cap \text{rad } P$. Then for any $e \in E(\overline{T_o})$, either $P \subseteq C_G^r(e)$ or else $P \subseteq C_G^{\ell}(e)$.

Proof. Let $\Delta \subseteq \Phi$ be the base relative to B. Then $-\Delta$ is the base relative to B^-. See Remark 4.48. If $a \in \Phi$, let U_α, $\sigma_\alpha: \alpha \to -\alpha$ be as in Definitions 4.43, 4.46. Then $\mathscr{S} = \mathscr{S}(B) = \mathscr{S}(B^-) = \{\sigma_\alpha | \alpha \in \Delta\}$. By Theorem 4.51, $P = BW_IB$ for some $I \subseteq \mathscr{S}$. Then $\mathscr{S} = I \cup \{\sigma_\gamma\}$ for some $\gamma \in \Delta$. Let $e \in E(\overline{T_o})$. Then $e^\sigma = e$ for all $\sigma \in W_I$. Now $P_1 = C_G^r(e)$ is a parabolic subgroup of G by Theorem 7.1. Clearly $T \subseteq P_1$, $W_I \subseteq W(P_1)$. By Remark 4.48, for each $\alpha \in \Delta$, either $U_\alpha \subseteq P_1$ or $U_{-\alpha} \subseteq P_1$. If $\alpha \neq \gamma$, then since $\sigma_\alpha \in W(P_1)$ and $\sigma_\alpha U_\alpha \sigma_\alpha^{-1} = U_{-\alpha}$, we see that both $U_\alpha, U_{-\alpha} \subseteq P_1$. Suppose $U_\gamma \subseteq P_1$. Let B_o denote the group generated by T, $U_\alpha (\alpha \in \Delta)$. Then B_o is a connected group by Proposition 4.2. Clearly $T \subseteq B_o \subseteq P_1 \cap B$. Now $B_o \subseteq B_1$ for some Borel subgroup $B_1 \subseteq P_1$. Since $U_\alpha \subseteq B_1$ for all $\alpha \in \Delta$, we see that each $\alpha \in \Delta$ is positive with respect to B_1. Thus $B = B_1 \subseteq P_1$. Since $W_I \subseteq W(P_1)$, $P = BW_IB \subseteq P_1$. Next suppose that $U_{-\gamma} \subseteq P_1$. Then $U_\alpha \subseteq P_1$ for all $\alpha \in -\Delta$. So as above, $B^- \subseteq P_1 = C_G^r(e)$. Thus $e \in \xi(B^-) = \xi^-(B)$. Hence $B \subseteq C_G^{\ell}(e)$. So $P = BW_IB \subseteq C_G^{\ell}(e)$. This proves the lemma.

Lemma 10.18. Suppose $M \subseteq \mathcal{M}_n(K)$ and that rad G consists of scalar matrices. Let $P \in \mathcal{P}(B)$, $P \neq G$. Then there exists $e \in E(\overline{T})$, $e \neq 0,1$ such that $P \subseteq C_G^r(e)$.

Proof. Let $G' = (G,G)$. Then G' is semisimple, $G = K^*G'$. Without loss of generality, we may assume that P is a maximal parabolic subgroup of G, $T \subseteq \mathcal{D}_n^*(K)$. Now $T = K^*T'$ for some maximal torus T' of G', $B = K^*B'$ for some Borel subgroup B' of G', $P = K^*P'$ for some maximal parabolic subgroup P' of G', $T' \subseteq B' \subseteq P'$. Let $T_o = T' \cap \text{rad } P'$. Then $\dim T_o = 1$. Let $E(\overline{KT_o}) = \{1,e,f,0\}$. Suppose $P \not\subseteq C_G^r(e)$, $P \not\subseteq C_G^r(f)$. Then there exist $a,b \in P'$ such that $ae \neq eae$, $bf \neq fbf$. Consider the representation $\phi: G' \to GL(n^2,K)$ given by $\phi(g) = g \otimes (g^{-1})^t$. Clearly the kernel of ϕ is a finite group of scalar matrices. Hence $\phi(P')$ is a maximal parabolic subgroup of $\phi(G')$ containing $\phi(T_o)$ in its radical. Let $M = \overline{K\phi(G')}$. Then $K^*\phi(P')$ is a maximal parabolic subgroup of $K^*\phi(G')$, containing $K^*\phi(T_o)$ in its radical. By Corollary 8.9, $E(\overline{K\phi(T_o)}) = \{1, e \otimes f, f \otimes e, 0\}$. Let $e_1 = e \otimes f$. Now $a_1 = a \otimes (a^{-1})^t$, $b_1 = b \otimes (b^{-1})^t \in \phi(P')$. Suppose $a_1e_1 = e_1a_1e_1$. Then

$$0 \neq ae \otimes (a^{-1})^t f = eae \otimes f(a^{-1})^t f$$

So $ae = \alpha eae$ for some $\alpha \in K^*$. Hence $ae = eae$, a contradiction. Thus $a_1e_1 \neq e_1a_1e_1$. Next suppose that $e_1b_1^{-1} = e_1b_1^{-1}e_1$. Then

$$0 \neq eb^{-1} \otimes fb^t = eb^{-1}e \otimes fb^t f$$

So $fb^t = \alpha fb^t f$ for some $\alpha \in K^*$. Hence $fb^t = fb^t f$ and $bf = fbf$, a contradiction. Thus $a_1e \neq e_1a_1e_1$, $eb_1^{-1} \neq e_1b_1^{-1}e_1$. This contradicts Lemma 10.17. Thus $P \subseteq C_G^r(e)$ or $P \subseteq C_G^r(f)$, proving the lemma.

Corollary 10.19. Let $P \in \mathcal{P}(B)$, $P \neq G$. Then $P \subseteq C_G^r(e)$ for some $e \in E(\overline{T})$, $e \neq 0,1$.

Proof. Without loss of generality, we may assume that P is maximal. Suppose the result is false. Then M has no central idempotents other than 0,1. By Corollary 6.31, rad G is one dimensional. By Corollary 6.13, there exists an idempotent separating homomorphism $\phi: M \to \mathcal{M}_n(K)$ such that the radical of $G' = \phi(G)$ consists of scalar matrices. By Theorem 10.12, $M' = \overline{\phi(M)} = \phi(M)$ and $\phi: \mathcal{E}(M) \cong \mathcal{E}(M')$. Since $\phi^{-1}(0) = \{0\}$ we see by Theorem 2.21 that $\phi^{-1}(1)$ is a finite group. Hence $\phi(P)$ is a maximal parabolic subgroup of G. By Lemma 10.18, there exists $e \in E(\overline{T})$, $e \neq 0,1$ such that $\phi(P) \subseteq C_{G'}^r(\phi(e)) \neq G$. Since $\phi(P)$ is maximal, $\phi(P) = C_{G'}^r(\phi(e))$. Then $\phi(e) \in \phi(\Lambda)$. So by Lemma 10.15, $P = C_G^r(e)$. This proves the result.

We are now in a position to prove the following result of the author [76].

Theorem 10.20. Let M be a connected regular monoid with zero and group of units G. Let T be a maximal torus of G, $\Lambda \in \mathcal{C}(T)$, $B = \beta(\Lambda)$. Let P be a parabolic subgroup of G containing B. Then there exists a chain $\Gamma \subseteq \Lambda$ such that $P = C_G^r(\Gamma)$.

Proof. We prove by induction on $\dim M$. We may assume that $P \neq G$. By Corollary 10.19, there exists $e \in \Lambda$ such that $P \subseteq C_G^r(e)$. There exists $I \subseteq \mathcal{S}$ such that $P = BW_IB$. Hence $e^\sigma = e$ for all $\sigma \in I$. By Lemma 10.16, $\Lambda_e = \{f \in \Lambda \mid e \leq f\} \in \mathcal{C}(T_e)$ in M_e and $e\Lambda = \{f \in \Lambda \mid e \geq f\} \in \mathcal{C}(eT)$ in eMe. Let $I_1 = \{\sigma \in I \mid \sigma_e = 1\}$, $I_2 = \{\sigma \in I \mid e\sigma = 1\}$, $I_1' = \{e\sigma \mid \sigma \in I_1\}$, $I_2' = \{\sigma_e \mid \sigma \in I_2\}$. Then by Lemma 10.16, $I = I_1 \cup I_2$, $I_1' \subseteq \mathcal{S}(e\Lambda) = \mathcal{S}_1$, $I_2' \subseteq \mathcal{S}(\Lambda_e) = \mathcal{S}_2$. So by Lemma 10.15 and the induction hypothesis applied to eMe and M_e, we see that there exist chains $\Gamma_1 \subseteq e\Lambda$, $\Gamma_2 \subseteq \Lambda_e$ such that $I_1' = C_{\mathcal{S}_1}(\Gamma_1)$, $I_2' = C_{\mathcal{S}_2}(\Gamma_2)$. Let $\Gamma = \Gamma_1 \cup \Gamma_2 \cup \{e\}$. Then Γ is a chain, $I \subseteq C_{\mathcal{S}}(\Gamma)$. Now let $\sigma \in C_{\mathcal{S}}(\Gamma)$. Then $e^\sigma = e$. By Lemma 10.16, either $\sigma_e = 1$, $e\sigma \in I_1'$ or else $e\sigma = 1$ and $\sigma_e \in I_2'$. By Proposition 10.9,

$\sigma \in I_1 \cup I_2 = I$. Hence $I = C_{\mathscr{S}}(\Gamma)$. By Lemma 10.15, $P = C_G^r(\Gamma)$, proving the theorem.

Corollary 10.21. Let M be a connected regular monoid with zero, $\Lambda \subseteq \mathscr{E}(M)$, a cross–section lattice. Then $\mathscr{S}(\Lambda) = \{\sigma \in W \mid \sigma \neq 1, \sigma$ fixes a chain of length ht $\mathscr{E}$ – 1 in $\Lambda\}$.

Proof. Let $B = \beta(\Lambda)$, ht $\mathscr{E} = m$. Let $\sigma \in W, \sigma \neq 1$. Suppose first that σ fixes a chain Γ of length $m - 1$ in Λ. Then $e^{\sigma} = e$ for all $e \in \Gamma$. Let $P = C_G^r(\Gamma) \supseteq B$. Then by Lemma 10.15, $\sigma \in W(P)$. Since $|\Gamma| = m - 1$, we see by Proposition 6.25 that $|W(P)| = 2$. Thus $P = B \cup B \sigma B$ and $\sigma \in \mathscr{S}(B)$. Conversely let $\sigma \in \mathscr{S}(B)$, $P = B \cup B\sigma B$. By Theorem 10.20, $P = C_G^r(\Gamma)$ for some chain $\Gamma \subseteq \Lambda$. Choose Γ such that $|\Gamma|$ is maximal. If there exists $e \in \Gamma, e \neq 0,1$, then proceeding inductively and using Lemmas 10.15, 10.16, we see that $|\Gamma| = m - 1$. If $\Gamma = \{0,1\}$, then $P = G$, $|W(G)| = 2$. By Theorem 10.10, σ fixes a chain Γ' in $\mathscr{E}$ of length $m - 1$. Then $w(e) = 1$ for all $e \in \Gamma'$. Hence Γ' consists of central idempotents by Corollary 6.31. So $\Gamma' \subseteq \Lambda$.

Let Λ be a cross–section lattice in a connected regular monoid M with zero.

Definition 10.22. (i) If $I \subseteq \mathscr{S}(\Lambda)$, let $\Lambda_I = \{e \in \Lambda \mid e^{\sigma} = e$ for all $\sigma \in I\}$, $\mathscr{U}_I = \{J_e \mid e \in \Lambda_I\}$. Let $\tilde{\mathscr{U}} = \{\mathscr{U}_I \mid I \subseteq \mathscr{S}(\Lambda)\}$.

(ii) If $\mathscr{V} \in \tilde{\mathscr{U}}$, let $\Lambda_{\mathscr{V}} = \{e \in \Lambda \mid J_e \in \mathscr{V}\}$.

Remark 10.23. (i) Since all cross–section lattices are conjugate, $\tilde{\mathscr{U}}$ is independent of Λ.

(ii) Let $I \subseteq \mathscr{S}(\Lambda)$, $B = \beta(\Lambda)$, $P = BW_I B$. Then $P \subseteq C_G^r(\Lambda_I)$. By Theorem 10.20, $P = C_G^r(\Gamma)$ for some chain $\Gamma \subseteq \Lambda$. Then $\Gamma \subseteq \Lambda_I$. Thus $P = C_G^r(\Lambda_I)$.

(iii) If $I, I' \subseteq \mathscr{S}(\Lambda)$, then $\mathscr{U}_I \cap \mathscr{U}_{I'} = \mathscr{U}_{I \cup I'}$. Also $I \subseteq I'$ if and only if $\mathscr{U}_{I'} \subseteq \mathscr{U}_I$.

(iv) $\tilde{\mathscr{U}}$ is a family of sublattices of $\mathscr{U}$, closed under intersection. With respect to inclusion $\tilde{\mathscr{U}}$ is a finite Boolean lattice. The importance of $\tilde{\mathscr{U}}$ will be clear in Chapter 14.

(v) In some cases, $\tilde{\mathscr{U}}$ can be determined directly from $\mathscr{U}$. For example, if $\mathscr{U}$ is a linear chain, then $\tilde{\mathscr{U}}$ is the family of all subchains of $\mathscr{U}$ containing $G, \{0\}$.

<u>Corollary 10.24</u>. If Λ is a cross–section lattice, $B = \beta(\Lambda)$, then the map $\phi: \tilde{\mathscr{U}} \to \mathscr{P}(B)$ given by $\phi(\mathscr{V}) = C_G^r(\Lambda_{\mathscr{V}})$ is an inclusion reversing bijection.

11 RENNER'S DECOMPOSITION AND RELATED FINITE SEMIGROUPS

In this chapter, we wish to generalize the Bruhat decomposition for groups (Theorem 4.35) to monoids. Some initial ideas in this direction were given by Grigor′ev [25]. However, the correct and complete solution is due to Renner [97]. Throughout this chapter let M be a connected regular monoid with zero 0 and group of units G. Fix a maximal torus T of G and let

$$\mathscr{E}_1 = \{e \in E(\overline{T}) \mid ht(e) = 1\}$$

denote the set of minimal non–zero 'diagonal idempotents' of M. We start with:

<u>Proposition 11.1</u>. $\overline{N_G(T)} = E(\overline{T})N_G(T)$ is a unit regular inverse monoid with group of units $N_G(T)$ and idempotent set $E(\overline{T})$. Moreover the fundamental congruence μ on $\overline{N_G(T)}$ is given by: $a \mu b$ if and only if $b \in Ta$.

<u>Proof</u>. Since $N_G(T)/T$ is a finite group, $\overline{N_G(T)} = N_G(T)\overline{T} = N_G(T)TE(\overline{T}) = N_G(T)E(\overline{T}) = E(\overline{T})N_G(T)$. Define θ on $\overline{N_G(T)}$ as: $a \theta b$ if $b \in Ta$. Then θ is an idempotent separating congruence on $\overline{N_G(T)}$. Hence $\theta \subseteq \mu$, Let $a,b \in \overline{N_G(T)}$, $a \mu b$. Then $a = ex$, $b = ey$ for some $e \in E(\overline{T})$, $x,y \in N_G(T)$. Let $z = xy^{-1}$. Then $e \mu exy^{-1}$. Hence $e \mathscr{H} exy^{-1}$. Let H denote the $\mathscr{H}$–class of e in M, $c = exy^{-1} \in H$. Then for all $f \in E(e\overline{T})$, $e \geq f$ and so $fc \mu f \mu cf$. Hence $fc = cf$.

Thus $c \in C_H(E(eT)) = eT$ by Theorem 7.1. So $exy^{-1} = et$ for some $t \in T$ and $a = ex = tey \in Tb$. Thus $\mu = \theta$.

Definition 11.2. The Renner monoid of M, Ren(M) is the fundamental inverse monoid, $\overline{N_G(T)}/\mu = \overline{N_G(T)}/T$.

Remark 11.3. (i) Ren(M) is a finite, unit regular inverse monoid with group of units $W = N_G(T)/T$ and idempotent set $E(\overline{T})$. Thus Ren(M) is the subsemigroup of the Munn semigroup of $E(\overline{T})$ generated by $E(\overline{T})$ and W. Thus Ren(M) can be determined from the $\mathscr{E}$–structure $\mathscr{E}(M)$ and conversely.

(ii) If $M = \mathscr{M}_n(K)$, then Ren(M) consists of the row and column monomial matrices in M with 0–1 entries and is thus isomorphic to the symmetric inverse semigroup of degree n.

The following result is due to Renner [97].

Theorem 11.4. Let B be a Borel subgroup of G containing T. Then M is the disjoint union of $BwB(w \in \text{Ren}(M))$.

Proof. Let M' denote the union of $BwB(w \in \text{Ren}(M))$. Then $BM'B = M'$. Let B^- denote the opposite Borel subgroup of G relative to T, $U = B_u$, $U^- = B^-_u$. We follow the notation of Definitions 4.43, 4.46, Remark 4.48. Let $\alpha \in \Delta(B)$, $\sigma = \sigma_\alpha$. So $W(G_\alpha) = \{1,\sigma\}$. Let $w \in \text{Ren}(M)$. Then $w = e\theta$ for some $e \in E(\overline{T})$, $\theta \in W$. By Corollary 4.53, we can find a closed connected subgroup Y of U such that $U = YU_\alpha = U_\alpha Y$, $\sigma Y = Y\sigma$. So $\sigma BwB = \sigma UTwB = \sigma UwTB = \sigma UwB$. Since $U = YU_\alpha$,

$$\sigma BwB = \sigma YU_\alpha wB = Y\sigma U_\alpha wB$$

By Corollary 9.11, either $U_\alpha e = \{e\}$ or $U_{-\alpha} = \{e\}$ or $e \in C(\overline{G}_\alpha)$. First suppose $U_\alpha e = \{e\}$. Then $\sigma BwB = Y\sigma U_\alpha e\theta B = Y\sigma e\theta B \subseteq M'$. Next suppose $e \in C(\overline{G}_\alpha)$. By Remark 4.48, either $U_\alpha \subseteq \theta B\theta^{-1}$ or else $U_{-\alpha} \subseteq \theta B\theta^{-1}$. First let $U_\alpha \subseteq \theta B\theta^{-1}$. Then $U_\alpha wB = eU_\alpha \theta B\theta^{-1}\theta = e\theta B\theta^{-1}\theta = e\theta B$. So

$$\sigma BwB = Y\sigma U_\alpha wB = Y\sigma e\theta B \subseteq M'$$

Next let $U_{-\alpha} \subseteq \theta B\theta^{-1}$. By Theorem 4.35 applied to G_α, we see that $G_\alpha = U_{-\alpha}T \cup U_{-\alpha}T\ \sigma\ U_{-\alpha}T$. Now $U_{-\alpha}\ TwB = U_{-\alpha}Te\theta B = eU_{-\alpha}T\theta B\theta^{-1}\theta = e\theta B\theta^{-1}\theta = e\theta B$. Thus

$$U_{-\alpha}TwB = e\theta B$$

So $G_\alpha wB \subseteq wB \cup U_{-\alpha}T\sigma wB = wB \cup U_{-\alpha}\sigma wB = wB \cup \sigma\ U_\alpha wB$. Since $\sigma G_\alpha = G_\alpha$, we see that $G_\alpha wB \subseteq \sigma wB \cup U_\alpha wB$. Hence

$$\sigma BwB = Y\sigma U_\alpha wB \subseteq YG_\alpha wB \subseteq Y\sigma wB \cup YU_\alpha wB \subseteq B\sigma wB \cup BwB \subseteq M'$$

So again $\sigma BwB \subseteq M'$. Finally suppose that $U_{-\alpha}e = \{e\}$. Then

$$U_{-\alpha}TwB = U_{-\alpha}Te\theta B = U_{-\alpha}eT\theta B = e\theta B$$

Hence we conclude as above that $\sigma BwB \subseteq M'$. Thus in all cases $\sigma BwB \subseteq M'$. Since W is generated by $\mathcal{S}(B)$, we see that $WBwB \subseteq M'$. Hence $WM' \subseteq M'$. Since $G = BWB$ by Theorem 4.35, $GM' \subseteq M'$. Similarly $M'G \subseteq M'$. Hence $M = GE(\overline{T})G \subseteq GM'G \subseteq M'$. Thus M is the union of $BwB(w \in Ren(M))$.

We now proceed to show that this union is disjoint. So let $a,a' \in \overline{N_G(T)}$ such that $BaB = Ba'B$. There exists $e,f \in E(\overline{T})$, $x,y \in N_G(T)$ such that $a = ex$, $a' = fy$. Then $fy = b_1exb_2$ for some $b_1,b_2 \in B$. Let J denote the $\mathcal{J}$-class of e. By Corollaries 3.20, 4.12, $eb_1e \in J$. Hence $ef \in J$. Thus $e = f$. Now $a = xx^{-1}ex$, $a' = yy^{-1}ey$. So the above argument shows that $x^{-1}ex = y^{-1}ey$. Thus $yx^{-1} \in C_G(e) \cap N_G(T)$. By Corollary 7.2, eBe, $exBx^{-1}e$ are Borel subgroups of the $\mathcal{H}$-class H of e. Now $eyx^{-1} \in N_H(eT)$, $eyx^{-1} = b_1exb_2x^{-1}$. Since $eyx^{-1} = yx^{-1}e$,

$$eyx^{-1} = eb_1exb_2x^{-1}e \in (eBe)(exBx^{-1}e)$$

By Theorem 4.35 applied to H, $eyx^{-1} \in eT$. Hence $ey \in exT$ and $a' \in aT$. This proves the theorem.

Corollary 11.5. $\overline{B}$ is the disjoint union of BwB, $w \in Ren(M) \cap \overline{B}$.

Remark 11.6. The finite monoid $Ren(M) \cap \overline{B}$ is only briefly encountered in [97]. It is a very interesting monoid, worthy of further study. It is a semilattice of nil semigroups.

The following is noted in the author [78].

Corollary 11.7. Let $J \in \mathcal{U}(M)$, $a \in J$. Then there exist $e,f \in E(\overline{T}) \cap J$ such that $eaf \in J$.

Proof. By Theorem 11.4, there exist $e \in E(\overline{T})$, $\sigma \in W$ such that $a \in Be\sigma B = B\sigma(\sigma^{-1}e\sigma)B$. By Corollary 3.20, $eBe \subseteq J$. Thus $ea \in J$. Similarly $af \in J$, where $f = \sigma^{-1}e\sigma$. By Theorem 1.4, $eaf \in J$.

As another application of Theorem 11.4, we obtain some maximal completely simple subsemigroups of M. This generalizes an informal conjecture to the author by Francis Pastijn concerning $\mathcal{T}_n(K)$.

Corollary 11.8. Let B be a Borel subgroup of G, $J_o \in \mathcal{U}(\overline{B})$. Then $\hat{J}_o = \{a \in M | a \mathcal{H} x$ for some $x \in J_o\}$ is a maximal completely simple subsemigroup of M.

Proof. Let $E_o = E(J_o)$. By Corollary 3.20, J_o is a completely simple semigroup. Let $a,b \in \hat{J}_o$. Then $a \mathcal{H} e$, $b \mathcal{H} e'$ for some $e,e' \in E_o$. Now $ee' \mathcal{H} f$ for some $f \in E_o$. Let J denote the $\mathcal{J}$-class of e in M. Now ee', $b = e'b \in J$. So by Theorem 1.4, $eb = ee'b \in J$. Now $a = ae \in J$. Thus by Theorem 1.4, $ab = aeb \in J$. Since $ea = a$, $be' = b$, we see by Theorem 1.4 that $ab \mathcal{R} e \mathcal{R} f$, $ab \mathcal{L} e' \mathcal{L} f$. Hence $ab \mathcal{H} f$ and $ab \in \hat{J}_o$. Thus $\hat{J}_o$ is a subsemigroup of M. Clearly $\hat{J}_o$ is completely regular and $E(\hat{J}_o) = E_o$. Thus $\hat{J}_o$ is a completely simple semigroup.

Now let S be a completely simple subsemigroup of M containing $\hat{J}_o$. We need to show that $E(S) = E_o$. So let $f \in E(S)$. Let T be a maximal torus of B. By Theorem 11.4, there exist $b_1,b_2 \in B$, $e \in E(\overline{T})$, $u \in N_G(T)$ such that $f = b_1eub_2$. Now $fE_o \cup E_of \subseteq S \subseteq J$, where J is the $\mathcal{J}$-class of e in M. Let $e_1 \in E_o \cap \overline{T}$. Then $b_1e_1b_1^{-1} \in E_o$, $b_1e_1b_1^{-1}f \in J$. Thus $e_1e \in J$. Hence $e_1 = e$. Similarly since $eu = uu^{-1}eu$, we see that $e_1 = u^{-1}eu$. Hence $e \in E_o$, $eu = ue$. Now $f^2 = f$. So $e = eub_2b_1e$ and $eu^{-1} = eb_2b_1e \in eBe$. By Corollary 7.2, eBe is a Borel subgroup of the $\mathcal{H}$-class H of e and $eu^{-1} \in N_H(eT)$. It follows that $eu^{-1} = e$. So $f = b_1eb_2 \in J_o$. Hence $f \in E_o$, completing the proof.

Problem 11.9. Let $J \in \mathcal{U}(M)$. Is every maximal closed, irreducible subsemigroup of J of the form described in Proposition 11.8?

Remark 11.10. Renner [97] obtains several interesting consequences of his decomposition. We mention a few.

(i) if $e \in \Lambda = \xi(B) = \xi^-(B^-)$, then B^-eB is open and dense in the $\mathcal{J}$-class $J = GeG$ of e. This generalizes to J, the big cell (Chevalley's) B^-B of

G.

(ii) The row echelon form for $\mathcal{M}_n(K)$ can be generalized to M. If $x \in Ren(M)$, then it is in 'row–echelon form' if $Bx \subseteq xB$.

For $\mathcal{M}_n(K)$, $\overline{N_G(T)}$ is the semigroup of all row and column monomial matrices. We now present analogues of the row monomial semigroup and the column monomial semigroup (due to the author [78]) for any connected regular monoid M with zero. Recall that in this chapter $\mathcal{E}_1$ denotes the set of minimal non–zero idempotents of T.

Definition 11.11. Let M be a connected regular monoid with zero.

(i) $\mathcal{Z}_r = \mathcal{Z}_r(M) = \{a \in M \mid \text{for all } e \in \mathcal{E}_1, \text{ there exists } f \in \mathcal{E}_1 \text{ such that } ea = eaf\}$.

(ii) $\mathcal{Z}_\ell = \mathcal{Z}_\ell(M) = \{a \in M \mid \text{for all } e \in \mathcal{E}_1, \text{ there exists } f \in \mathcal{E}_1 \text{ such that } ae = fae\}$.

(iii) $\mathcal{Z}'_r = \mathcal{Z}'_r(M) = \{a \in M \mid aT \subseteq Ta\}$.

(iv) $\mathcal{Z}'_\ell = \mathcal{Z}'_\ell(M) = \{a \in M \mid Ta \subseteq aT\}$.

The following result is due to the author [78].

Theorem 11.12. Let M be a connected regular monoid with zero 0. Then

(i) $\mathcal{Z}_r$, $\mathcal{Z}'_r$, $\mathcal{Z}_\ell$, $\mathcal{Z}'_\ell$ are unit regular submonoids of M with group of units $N_G(T)$.

(ii) $\mathcal{Z}'_\ell \subseteq \mathcal{Z}_r$, $\mathcal{Z}'_\ell \subseteq \mathcal{Z}_\ell$, $\mathcal{Z}_r \cap \mathcal{Z}_\ell = \mathcal{Z}'_r \cap \mathcal{Z}'_\ell = \overline{N_G(T)}$.

(iii) For all $a \in \mathcal{Z}_r$, there exists $f \in E(\overline{T})$ such that $a \mathcal{L} f$ in $\mathcal{Z}_r$. For all $a \in \mathcal{Z}_\ell$, there exists $f \in E(\overline{T})$ such that $a \mathcal{R} f$ in $\mathcal{Z}_\ell$.

(iv) For all $e \in E(\overline{T})$, $\mathcal{Z}_r(eMe) = \mathcal{Z}_r(M) \cap eMe$, $\mathcal{Z}_\ell(eMe) = \mathcal{Z}_\ell(M) \cap eMe$.

(v) If $e \in \mathcal{E}_1$, then for all $e' \in E(M)$, $e \mathcal{L} e'$ implies $e' \in \mathcal{Z}'_r$, $e \mathcal{R} e'$ implies $e' \in \mathcal{Z}'_\ell$.

<u>Proof</u>. We first prove (iv). Let H denote $\mathcal{H}$–class of e. So eT is a maximal torus of H by Theorem 6.16. Let $\mathcal{E}'_1$ be the set of minimal elements of $E(eT)\backslash\{0\}$. So $e\mathcal{E}_1 = \mathcal{E}'_1 \cup \{0\}$. Let $a \in \mathcal{Z}_r(eMe)$, $e_1 \in \mathcal{E}_1$. If $e_1 \notin \mathcal{E}'_1$, then $e_1a = e_1ea = 0 = e_1ae_1$. If $e_1 \in \mathcal{E}'_1$, then $e_1a = e_1af_1$ for some $f_1 \in \mathcal{E}'_1 \subseteq \mathcal{E}_1$. So $a \in \mathcal{Z}_r(M) \cap eMe$. Next let $a \in \mathcal{Z}_r(M) \cap eMe$. Let $e_1 \in \mathcal{E}'_1$. Then $e_1a = e_1af$ for some $f \in \mathcal{E}_1$. Let $f_1 = ef \in \mathcal{E}'_1 \cup \{0\}$. Then $e_1a = e_1af_1$ and $a \in \mathcal{Z}_r(eMe)$. Similarly $\mathcal{Z}_\ell(eMe) = \mathcal{Z}_\ell(M) \cap eMe$.

It is easily checked that $\mathcal{Z}_r$, $\mathcal{Z}'_r$, $\mathcal{Z}_\ell$, $\mathcal{Z}'_\ell$ are submonoids of M and that $\overline{N_G(T)} = E(T)N_G(T) \subseteq \mathcal{Z}'_r \cap \mathcal{Z}'_\ell$. Let $a \in \mathcal{Z}'_r$, $e \in \mathcal{E}_1$. Then $ea \in \mathcal{Z}'_r$. Suppose $ea \neq 0$. Then $ea \mathcal{R} e$ in M. So by Proposition 6.1, $ea = ex$ for some $x \in G$. Now $ex\,T \subseteq Tex$. So $exTx^{-1} \subseteq Te$. Thus $exTx^{-1} = exTx^{-1}e$. Hence T, xTx^{-1} are maximal tori of $C^\ell_G(e)$. So there exists $u \in C^\ell_G(e)$ such that $T = uxTx^{-1}u^{-1}$. Thus $ux \in N_G(T)$. Hence $f = x^{-1}u^{-1}eux \in \mathcal{E}_1$. Now $eaf = exf = exx^{-1}u^{-1}eux = eu^{-1}eux = eu^{-1}ux = ex = ea$. Thus $a \in \mathcal{Z}_r$. Hence $\mathcal{Z}'_r \subseteq \mathcal{Z}_r$. Similarly $\mathcal{Z}'_\ell \subseteq \mathcal{Z}_\ell$.

We show next that $N_G(T) = G \cap \mathcal{Z}_r$. So let $x \in G \cap \mathcal{Z}_r$. Let $\mathcal{E}_1(x) = \{e \in \mathcal{E}_1 \mid ex = exe\}$, $\alpha(x) = |\mathcal{E}_1| - |\mathcal{E}_1(x)|$. We prove by induction on $\alpha(x)$ that $x \in N_G(T)$. Suppose first that $\alpha(x) = 0$. Then $\mathcal{E}_1(x) = \mathcal{E}_1$ and by Proposition 7.5, $x \in C^\ell_G(E(T))$. By Theorem 9.10, $x \in T \subseteq \mathcal{Z}_r$. So assume $\alpha(x) > 0$. Then $x \in C^\ell_G(\mathcal{E}_1(x))$. Let $e \in \mathcal{E}_1\backslash\mathcal{E}_1(x)$. Then $ex = exf$ for some $f \in \mathcal{E}_1$. So $f \mid e$ in $M' = \overline{C^\ell_G(\mathcal{E}_1(x))}$. Then $e \mathcal{J} f$ in M'. By Proposition 6.25, there exists $u \in N_G(T) \cap C^\ell_G(\mathcal{E}_1(x))$ such that $u^{-1}eu = f$. So $ex = exf = exu^{-1}eu$. Thus $exu^{-1} = exu^{-1}e$. Hence $\mathcal{E}_1(xu^{-1}) \supseteq \mathcal{E}_1(x) \cup \{e\}$ and $\alpha(xu^{-1}) < \alpha(x)$. Thus $xu^{-1} \in N_G(T)$, whereby $x \in N_G(T)$. Thus $N_G(T) = G \cap \mathcal{Z}_r = G \cap \mathcal{Z}'_r$. Similarly $N_G(T) = G \cap \mathcal{Z}_\ell =$

$G \cap \mathscr{Z}'_{\ell}$.

Next we prove the following claim:

$$e \in E(\overline{T}),\ a \in \mathscr{Z}_r,\ eae\ \mathscr{H}\ e \text{ in } M \text{ imply } ea = eae \tag{10}$$

We may assume that $e \neq 0$. Let $\Gamma = \{f \in \mathscr{E}_1 \,|\, f \leq e\} = \{f_1,...,f_k\}$. Since $E(e\overline{T})$ is relatively complemented, we see that $e = f_1 \vee ... \vee f_k$. Let $a \in \mathscr{Z}_r$ such that $eae\ \mathscr{H}\ e$ in M. Let $\Gamma(a) = \{f \in \Gamma \,|\, fa = faf\}$, $\gamma(a) = |\Gamma| - |\Gamma(a)|$. We prove by induction on $\gamma(a)$ that $ea = eae$. First suppose that $\gamma(a) = 0$. Then $fa = faf$ for all $f \in \Gamma$. Let $f \in \Gamma$. Then $faf = fa = f(ea) \,|\, f(eae) \,|\, fe = f$. So $faf\ \mathscr{H}\ f$ for all $f \in \Gamma$. By Corollary 6.17, $a \in \overline{C^{\ell}_G(\Gamma)}$. By Proposition 7.5, $C^{\ell}_G(\Gamma) \subseteq C^{\ell}_G(e)$. Thus $ea = eae$. So assume $\gamma(a) > 0$. As above, $faf\ \mathscr{H} f$ for all $f \in \Gamma(a)$. Let $M' = \overline{C^{\ell}_G(\Gamma(a))}$. By Corollary 6.17, $a \in M'$. By Remark 1.3, $eae\ \mathscr{H} e$ in M' and $faf\ \mathscr{H} f$ in M' for all $f \in \Gamma(a)$. There exists $f_i \in \Gamma\backslash\Gamma(a)$. There exists $f'_i \in \mathscr{E}_1$ such that $f_ia = f_iaf'_i$. Now in M', $f'_i \,|\, f_ia = f_i(ea) \,|\, f_i(eae) \,|\, f_ie = f_i$. Thus $f_i\ \mathscr{J}\ f'_i$ in M' and $f_iae = f_ieae \neq 0$. Suppose $f'_i \notin \Gamma$. Then $f'_ie = 0$ and $f_iae = f_iaf'_ie = 0$, a contradiction. So $f'_i \in \Gamma$. By Proposition 6.25, 6.27, there exists $y \in N_G(T) \cap C^{\ell}_G(\Gamma(a)) \cap C_G(e)$ such that $yf_iy^{-1} = f'_i$. So $f_ia = f_iaf'_i = f_iayf_iy^{-1}$. Thus $f_iay = f_iayf_i$. Clearly $fay = fayf$ for all $f \in \Gamma(a)$. Also $eaye = eaey = (eae)(ey)\ \mathscr{H} e$. Clearly $\Gamma(ay) \supseteq \Gamma(a) \cup \{f_i\}$. Hence $\gamma(ay) < \gamma(a)$ and $eay = eaye$. Since $y \in C_G(e)$, $ea = eae$. This proves claim (10).

Now let $a \in \mathscr{Z}_r$, J the $\mathscr{J}$-class of a in M. By Corollary 11.7, there exist $e,f \in J \cap E(\overline{T})$ such that $fae \in J$. By Proposition 6.25, $f = x^{-1}ex$ for some $x \in N_G(T)$. Then $exae \in J$. By Theorem 1.4, $exae\ \mathscr{H}\ e$. By (10), $exa = exae$. By Theorem 1.4, $xa\ \mathscr{L}\ exa = exae$. So $xae = xa$ and $ae = a$. Let H denote the $\mathscr{H}$-class of e. Then $exa = exae \in H \cap \mathscr{Z}_r(M) = H \cap \mathscr{Z}_r(eMe) = N_H(eT)$. By Theorem 6.16, there exists $u \in C_G(e) \cap N_G(T)$ such that $eu = ue = exa$. So

$e = eu^{-1}xa$. Let $e_o = u^{-1}xa$. Since $ae = a$, we see that $e_o \in E(\mathcal{Z}_r)$. Clearly $e \mathcal{L} e_o$. Since $u^{-1}x \in N_G(T)$, we see that $\mathcal{Z}_r$ is unit regular. Since $\mathcal{Z}'_r \subseteq \mathcal{Z}_r$ and has $N_G(T)$ as the group of units, we see that $\mathcal{Z}'_r$ is also unit regular. Similarly $\mathcal{Z}'_\ell$, $\mathcal{Z}_\ell$ are the unit regular. Since $a \mathcal{L} e$ in $\mathcal{Z}_r$, we see that (iii) is also valid. We have thus proved (i), (iii).

Since $N_G(T)$ is the group of units of $\mathcal{Z}_r \cap \mathcal{Z}'_r$, we see that $\mathcal{Z}_r \cap \mathcal{Z}_\ell$ is also unit regular. Let $h \in E(\mathcal{Z}_r \cap \mathcal{Z}_\ell)$. Then by (iii), there exist $e,f \in E(\overline{T})$ such that $e \mathcal{L} h \mathcal{R} f$. By Theorem 1.4, $e \mathcal{J} ef = fe$. It follows that $e = f = h$. Hence $E(\mathcal{Z}_r \cap \mathcal{Z}_\ell) = E(\overline{T})$. Thus $\mathcal{Z}_r \cap \mathcal{Z}_\ell = \mathcal{Z}'_r \cap \mathcal{Z}'_\ell = \overline{N_G(T)}$. This proves (ii).

Let $e \in \mathcal{E}_1$. Then by Proposition 6.2, $\dim eMe = 1$. Let $V = \operatorname{rad} G$, H the $\mathcal{H}$-class of e in M. Then $eV \subseteq H$, $0 \in \overline{V}$. Thus $eV = H$. Now let $e' \in E(M)$, $e \mathcal{L} e'$. Let H' denote the $\mathcal{H}$-class of e' in M. By Corollary 6.19, $C_G^\ell(e) = C_G^\ell(e')$. Thus $T \subseteq C_G^\ell(e')$. So $e'T = e'Te' \subseteq H' = e'V = Ve' \subseteq Te'$. Thus $e' \in \mathcal{Z}'_r$. This proves (vi).

Finally we prove (v) by induction on $\dim M$. Thus we are reduced to the case when $e \in \mathcal{E}_1$. Let $V = \operatorname{rad} G$. Then $VT_e \subseteq T$, $0 \in \overline{V}$, $\dim T_e = \dim T - 1$. Thus $T = VT_e$. Let $a \in \mathcal{Z}'_r(M_e)$. Then $aT_e \subseteq T_e a$. Since $T = VT_e$, $aT \subseteq Ta$ and $a \in \mathcal{Z}_r(M)$. This proves the theorem.

Renner [91; Theorem 4.4.4] has shown that $C_M(T) = \overline{T}$. We now generalize this result.

Corollary 11.13. Let M be a connected regular monoid with zero. Then $C_M(E(\overline{T})) = \overline{T} = C_M(\mathcal{E}_1)$.

Proof. By Theorem 11.12, $\overline{T} \subseteq C_M(E(\overline{T})) \subseteq C_M(\mathcal{E}_1) \subseteq \mathcal{Z}_r \cap \mathcal{Z}_\ell \cap C_M(\mathcal{E}_1) = \overline{N_G(T)} \cap C_M(\mathcal{E}_1) = \overline{T}$.

Example 11.14. Let $M = \mathcal{M}_n(K)$. Then $\mathscr{Z}_r = \mathscr{Z}'_r$ consists of all row monomial matrices and $\mathscr{Z}_\ell = \mathscr{Z}'_\ell$ consists of all column monomial matrices.

Example 11.15. Let $M = \{A \oplus B \mid A,B \in \mathcal{M}_2(K),\ \det A = \det B\}$, G the group of units of M. Then $\mathscr{Z}_r \backslash G = Y \times Y$ where $Y = \left\{\begin{bmatrix} x & 0 \\ y & 0 \end{bmatrix} \middle| x,y \in K\right\} \cup \left\{\begin{bmatrix} 0 & x \\ 0 & y \end{bmatrix} \middle| x,y \in K\right\}$. However it is easy to see that $\begin{bmatrix} 1 & 0 \\ 1 & 0 \end{bmatrix} \oplus \begin{bmatrix} 1 & 0 \\ 1 & 0 \end{bmatrix} \notin \mathscr{Z}'_r$. Thus $\mathscr{Z}_r \neq \mathscr{Z}'_r$.

Remark 11.16. It follows easily from Theorem 11.12 that $\mathscr{U}(\mathscr{Z}_r) \cong \mathscr{U}(\mathscr{Z}_\ell) \cong \mathscr{U}(M)$, $\mathscr{Z}_r/\mathscr{L} \cong \mathscr{Z}_\ell/\mathscr{R} \cong E(\overline{T})$.

Remark 11.17. The inverse semigroup $\overline{N_G(T)}$ acts on $E(\overline{T})$ on the right as follows: for $e \in E(\overline{T})$, $a \in \overline{N_G(T)}$, let $e \cdot a = a^{-1}ea$. This action extends to $\mathscr{Z}_r$ as follows: for $e \in E(\overline{T})$, $a \in \mathscr{Z}_r$, let $e \cdot a = f$ where $f \in E(\overline{T})$ is such that $ea \mathscr{L} f$ in $\mathscr{Z}_r$. The action of $\overline{N_G(T)}$ on $E(\overline{T})$ gives rise to the finite inverse monoid Ren(M). Similarly the action of $\mathscr{Z}_r$ on $E(\overline{T})$ gives rise to a finite fundamental regular monoid $\mathscr{Y}_r$. We will obtain this monoid in the next theorem in a slightly different way.

If X is a set then we let $\mathscr{P}\mathscr{T}(X)$ denote the regular semigroup of all partial transformations on X and $\mathscr{I}(X)$ the inverse semigroup of all partial one to one transformations on X. The action is always on the right. See [33; Section I.4]. The following result is from the author [78].

Theorem 11.18. Let M be a connected regular monoid with zero 0. For $a \in \mathscr{Z}_r$, define $\pi(a) \in \mathscr{P}\mathscr{T}(\mathscr{E}_1)$ as follows: if $e,f \in \mathscr{E}_1$, then $e\,\pi(a) = f$ if $ea = eaf \neq 0$, $e\pi(a)$ is undefined if $ea = 0$. Then

(i) π: $\mathscr{Z}_r \to \mathscr{P}\mathscr{T}(\mathscr{E}_1)$ is a well defined homomorphism which is one to one on $E(\overline{T})$. Moreover $\pi(ta) = \pi(a)$ for all $t \in T$.

(ii) The kernel π^* of π is the largest congruence on $\mathscr{Z}_r$ contained in $\mathscr{L}$.

(iii) π^* restricted to $\overline{N_G(T)}$ is the fundamental congruence on $\overline{N_G(T)}$ and hence $\mathscr{Y} = \pi(\overline{N_G(T)}) \cong \mathrm{Ren}(M)$.

(iv) $\mathscr{Y}_r = \pi(\mathscr{Z}_r)$ is a finite fundamental unit regular monoid having no non–trivial congruences contained in $\mathscr{L}$

(v) $\mathscr{Y} = \mathscr{Y}_r \cap \mathscr{I}(\mathscr{E}_1)$.

Proof. (i) Since $f_1f_2 = 0$ for all $f_1, f_1 \in \mathscr{E}_1$ with $f_1 \neq f_2$, π is well defined. Let $a,b \in M$, $e \in \mathscr{E}_1$. First suppose that $eab \neq 0$. Then $ea \neq 0$, there exists $f \in \mathscr{E}_1$ such that $ea = eaf$. So $e\pi(a) = f$. If $fb = 0$, then $eab = eafb = 0$, a contradiction. So $fb \neq 0$. Let $f\pi(b) = f' \in \mathscr{E}_1$. So $fb = fbf'$. Thus $e\pi(a)\pi(b) = f'$. Also, $eab = eafb = eafbf' = eabf'$ and $e\pi(ab) = f'$. Next suppose that $\pi(a)\,\pi(b)$ is defined on e and $e\pi(a) = f$, $f\pi(b) = f'$. So $ea = eaf$, $fb = fbf'$, $ea \neq 0$, $fb \neq 0$. So $fb\ \mathscr{R}\ f$ in M. Now $eab = eafb = eafbf' = eabf'$. If $eab = 0$, then $eafb = 0$, implying $eaf = 0$. Hence $ea = eaf = 0$, a contradiction. Thus $eab \neq 0$ and $e\pi(ab) = f'$. It follows that $\pi(ab) = \pi(a)\pi(b)$. For $e \in E(\overline{T})$, set $\mathscr{E}_1(e) = \{e' \in \mathscr{E}_1 \mid e' \leq e\}$. Clearly $\mathscr{E}_1(e)$ is the domain of $\pi(e)$. Moreover, since $E(\overline{T})$ is relatively complemented, e is the join of $\mathscr{E}_1(e)$ in $E(\overline{T})$. Thus the map, $e \to \mathscr{E}_1(e)$ is injective. Hence π is 1–1 on $E(\overline{T})$. That $\pi(ta) = \pi(a)$ for $a \in \mathscr{Z}_r$, $t \in T$, is obvious.

(ii) By (i) and Theorem 11.12, $\pi^* \subseteq \mathscr{L}$. Let δ be a congruence on $\mathscr{Z}_r$ such that $\delta \subseteq \mathscr{L}$. Let $a,b \in \mathscr{Z}_r$ such that $a\ \delta\ b$. Let $e \in \mathscr{E}_1$. Then $ea\ \mathscr{L}\ eb$. So $ea \neq 0$ if and only if $eb \neq 0$. So suppose $ea \neq 0$, $eb \neq 0$. Let $e\,\pi(a) = f$, $e\,\pi(b) = f'$. Then $ea = eaf \neq 0$, $eb = ebf' \neq 0$. So in M, $f\ \mathscr{L}\ ea\ \mathscr{L}\ eb\ \mathscr{L}\ f'$. Hence $f = f'$ and $\pi(a) = \pi(b)$. Thus $\delta \subseteq \pi^*$.

(iii) follows from (i) and Proposition 11.1. (iv) follows from (ii). So we need only prove (v). Now $\mathscr{Y}$ and $\mathscr{Y}_r \cap \mathscr{I}(\mathscr{E}_1)$ have the same group of units W. It is also clear that $\mathscr{Y} \subseteq \mathscr{Y}_r \cap \mathscr{I}(\mathscr{E}_1)$. So it suffices to show that $E(\mathscr{Y}_r \cap \mathscr{I}(\mathscr{E}_1)) \subseteq$

$E(\mathscr{Y})$. Let $h \in E(\mathscr{Y}_r)$ such that $\pi(h) \in \mathscr{I}(\mathscr{E}_1)$. By Theorem 11.12, there exists $e \in E(T)$ such that $e \mathscr{L} h$. Then $\pi(e), \pi(h) \in \mathscr{I}(\mathscr{E}_1)$, $\pi(e) \mathscr{L} \pi(h)$. Since $\mathscr{I}(\mathscr{E}_1)$ is an inverse semigroup, $\pi(h) = \pi(e) \in \mathscr{Y}$. This proves the theorem.

Remark 11.19. If $M = \mathscr{M}_n(K)$, then $\mathscr{Y}_r = \mathscr{P}\mathscr{I}(\mathscr{E}_1)$, $\mathrm{Ren}(M) \cong \mathscr{Y} = \mathscr{I}(\mathscr{E}_1)$.

Remark 11.20. Let $\mathscr{Y}_r' = \pi(\mathscr{Z}_r')$. Then $\mathscr{Y}_r'$ has W as the group of units and hence is unit regular. Example 11.15 shows that in general $\mathscr{Y}_r \neq \mathscr{Y}_r'$. There is a natural congruence which can be defined on $\mathscr{Z}_r'$: $a \,\theta\, b$ if $b \in Ta$. By Theorem 11.18, $\theta \subseteq \pi^*$. For $\mathscr{M}_n(K)$, $\theta = \pi^*$. In general $\mathscr{Z}_r'/\theta$ need not even be a finite semigroup. To see this let $M = \overline{KG_o} \subseteq \mathscr{M}_9(K)$, where $G_o = \{A \otimes (A^{-1})^t \mid A \in SL(3,K)\}$. Let $X = \left\{ \begin{bmatrix} 1 & 0 & 0 \\ 1 & 0 & 0 \\ 1 & 0 & 0 \end{bmatrix} \otimes \begin{bmatrix} 0 & \alpha & 0 \\ 0 & 1 & 0 \\ 0 & \gamma & 0 \end{bmatrix} \middle| \alpha, \gamma \in K, \alpha + \gamma = -1 \right\}$. By Example 8.5, $X \subseteq E(M)$. It is routinely verified that no two elements of X are θ–related.

Problem 11.21. Are the semigroups $\mathscr{Y}_r$ weakly inverse in the sense of Srinivasan [109]?

12 BIORDERED SETS

Let S be a regular semigroup. Then what structure does the system E(S) possess? The correct answer is provided by the following definition due to Nambooripad [51], [52].

Definition 12.1. Let E be a set with a partial binary operation. For e, $f \in E$ define $f \leq_r e$ if $ef = f$, $f \leq_\ell e$ if $fe = f$. Let $\leq = \leq_r \cap \leq_\ell$, $\mathscr{R} = \leq_r \cap (\leq_r)^{-1}$, $\mathscr{L} = \leq_\ell \cap (\leq_\ell)^{-1}$. Suppose that the following conditions and their right–left duals hold for $e,f,g \in E$.

(i) $\leq_r$, $\leq_\ell$ are quasi–orders on E and ef is defined if and only if $e \leq_r f$ or $e \leq_\ell f$ or $f \leq_r e$ or $f \leq_\ell e$.

(ii) $f \leq_r e$ implies $f \,\mathscr{R}\, fe \leq e$.

(iii) $f \leq_r e$, $g \leq_r e$, $g \leq_\ell f$ imply $ge \leq_\ell fe$ and $(fe)(ge) = (fg)e$

(iv) $g \leq_r f \leq_r e$ implies $gf = (ge)f$.

For $e,f \in E$, let $fE = \{g \in E \mid g \leq_r f\}$, $Ee = \{g \in E \mid g \leq_\ell e\}$, $fEe = fE \cap Ee$. For $g,h \in fEe$, define $g \preceq h$ if $eg \leq_r eh$ and $gf \leq_\ell hf$. Let the sandwich set, sand(e,f) = $\{h \in fEe \mid g \preceq h$ for all $g \in fEe\}$. Assume that the following condition and its right–left dual holds for all $e,f,g \in E$:

(v) $f \leq_r e$, $g \leq_r e$ imply sand (f,g)e = sand(fe,ge).

Then E is called a biordered set. E is a regular biordered set if further,

(vi) sand(e,f) $\neq \emptyset$ for all $e,f \in E$.

Remark 12.2 (i) If $e \mathscr{R} f$ then $e \in \text{sand}(e,f)$. If $e \mathscr{L} f$, then $f \in \text{sand}(e,f)$.

(ii) Let S be a semigroup, $e,f \in E(S)$. If $ef = f$, define $e \circ f = f$, $f \circ e = fe \in E(S)$. If $fe = f$, then define $f \circ e = f$, $e \circ f = ef \in E(S)$. Then it is routinely verified that $E(S) = (E(S),\circ)$ is a biordered set. If S is regular, then $E(S)$ is a regular biordered set.

(iii) The basic theorem of Nambooripad [51], [52] is that every regular biordered set is the biordered set of idempotents of some regular semigroup. Recently David Easdown has shown that any biordered set is the biordered set of idempotents of some semigroup [20].

(iv) The axioms for a biordered set are quite complicated. However, considering the general nature of semigroups, it is rather surprising that such a finite axiomatization is even possible. For connected regular monoids with zero, the quasi-orders $\leq_r, \leq_\ell$ determine the biordered set E (Corollary 12.5). This is not so in general [52; Example 1.1].

(v) In the synthetic construction of biordered sets, one often starts with two quasi-orders $\leq_r, \leq_\ell$. Then, whenever $f \leq_r e$, one must define fe and whenever $f \leq_\ell e$, one must define ef. This partial binary operation would then first have to be shown to be well defined.

(vi) A regular biordered set with $\leq_r = \leq_\ell$ is just a semilattice.

Definition 12.3. Let E be a biordered set.

(i) If $e,f \in E$, then $e \mathscr{R}' f$ if for all $e_1 \in E$ with $e \mathscr{R} e_1$, there exists $f_1 \in E$ with $e_1 \mathscr{L} f_1 \mathscr{R} f$, and for all $f_1 \in E$ with $f_1 \mathscr{R} f$, there exists $e_1 \in E$ with $e \mathscr{R} e_1 \mathscr{L} f_1$. Similarly $e \mathscr{L}' f$ if for all $e_1 \in E$ with $e \mathscr{L} e_1$, there exists $f_1 \in E$ with $e_1 \mathscr{R} f_1 \mathscr{L} f$, and for all $f_1 \in E$ with $f_1 \mathscr{L} f$, there exists $e_1 \in E$ with $e \mathscr{L} e_1 \mathscr{R} f_1$. Clearly $\mathscr{R}'$, $\mathscr{L}'$ are equivalence relations on E and $\mathscr{R} \subseteq \mathscr{R}'$, $\mathscr{L} \subseteq \mathscr{L}'$.

(ii) E is reduced if $\mathscr{R} = \mathscr{R}'$, $\mathscr{L} = \mathscr{L}'$.

(iii) E is <u>locally reduced</u> if eEe is reduced for all $e \in E$.

(iv) A semigroup S is <u>reduced</u> if E(S) is reduced. S is <u>locally reduced</u> if E(S) is locally reduced, i.e. eSe is reduced for all $e \in E(S)$.

The following result is due to the author [86].

<u>Theorem 12.4</u>. Let E be a locally reduced biordered set, $e,f \in E$, $f \le_r e$. Then $f_1 = fe \in E$ is characterized within the system $(E, \le_r, \le_\ell)$ by the following threee properties.

(a) $f \mathcal{R} f_1 \le e$.

(b) For all $h \in E$ with $f_1 \mathcal{L} h \le e$, there exists $h' \in E$ such that $f \mathcal{L} h' \mathcal{R} h$.

(c) For all $h' \in E$ with $f \mathcal{L} h' \le_r e$, there exists $h \in E$ such that $h' \mathcal{R} h \mathcal{L} f_1$.

Similarly if $f \le_\ell e$, then ef is characterized within the system $(E, \le_r, \le_\ell)$ by the dual conditions. In particular, the system $(E, \le_r, \le_\ell)$ uniquely determines the biordered set E.

<u>Proof</u>. Let $e,f \in E$, $f \le_r e$. Then by Definition 12.1 (ii), $f \mathcal{R} fe \le e$. Let $h \in E$ with $fe \mathcal{L} h \le e$. Then by 12.1 (v) and Remark 12.2 (i), $h \in \text{sand}(fe,h) = \text{sand}(f,h)\,e$. So there exists $h' \in \text{sand}(f,h)$ such that $h = h'e$. By 12.1 (ii), $h \mathcal{R} h'$. Now $h' \le_\ell f$, $h' \le_r e$, $f \le_r e$. So by 12.1 (iii), $(fh')e = (fe)(h'e) = (fe)h = fe$. Hence $f \mathcal{R} fe \mathcal{R} fh' \le f$. Thus $fh' = f$ and $f \mathcal{L} h'$. So fe satisfies (a), (b). Now let $h' \in E$ with $f \mathcal{L} h' \le_r e$. Let $h = h'e \le e$. Then $h \mathcal{R} h'$ by 12.1 (ii). By 12.1 (iii), $fe \mathcal{L} h'e = h$. Hence fe satisfies (a), (b), (c).

Finally let $f_1, f_2 \in E$, satisfying (a), (b), (c). Then by (a), $f_1,f_2 \in eEe$, $f_1 \mathcal{R} f_2$. Let $h_1 \in eEe$ such that $f_1 \mathcal{L} h_1$. Then by (b), there exists $h' \in E$ such that $f \mathcal{L} h' \mathcal{R} h$. So $h' \le_r e$. By (c), there exists $h_2 \in E$ with $h' \mathcal{R} h_2 \mathcal{L} f_2$. Since $f_2 \le e$, $h_2 \le_\ell e$, $h_1 \mathcal{R} h_2 \mathcal{L} f_2$. It follows that $f_1 \mathcal{L}' f_2$ in eEe. Since E is locally reduced, $f_1 \mathcal{L} f_2$. Thus $f_1 = f_2$.

Corollary 12.5. Let M be a connected regular monoid with zero. Then M is locally reduced and hence the biordered set E(M) is completely determined by the system $(E(M), \leq_r, \leq_\ell)$.

Proof. For $e \in E(M)$, eMe is also a connected regular monoid with zero. So it suffices to show that M is reduced. Let G denote the group of units of M. Let $e,f \in E(M)$ such that $e \mathcal{R}' f$. By Proposition 7.6, there exists a maximal torus T of G, $e_1,f_1 \in E(\overline{T})$ such that $e \mathcal{R} e_1$, $f \mathcal{R} f_1$. There exists $h \in E(M)$ such that $e_1 \mathcal{L} h \mathcal{R} f$. Since $f \mathcal{R} f_1$, we see by Theorem 1.4 (vi) that $e_1 \mathcal{R} f_1e_1$. But $f_1e_1 = e_1f_1$. So by Theorem 1.4 (i), $e_1 = f_1$. Hence $e \mathcal{R} f$ and $\mathcal{R} = \mathcal{R}'$. Similarly $\mathcal{L} = \mathcal{L}'$.

Definition 12.6. Let E be a regular biordered set, $\mathcal{N}'(E) = \{(e,\psi,f) \mid e,f \in E, \psi\colon eEe \to fEf$ is an isomorphism$\}$. Define $(e,\psi,f) \equiv (e',\psi',f')$ if $e \mathcal{R} e'$, $f \mathcal{L} f'$, $f'\psi(x) = \psi'(xe')$ for all $x \in eEe$. Then $\equiv$ is an equivalence relation on $\mathcal{N}'(E)$. For $(e,\psi,f) \in \mathcal{N}'(E)$, let $[e,\psi,f]$ denote its $\equiv$-class. Let $\mathcal{N}(E) = \mathcal{N}'(E)/\equiv$. If $a = [e,\psi,f]$, $b = [e',\psi',f'] \in \mathcal{N}(E)$, $h \in$ sand (f,e'), then let

$$ab = [\psi^{-1}(fh), \lambda, \psi'(he')]$$

where $\lambda(x) = \psi'(h\psi(x))e')$. This is a well-defined associative binary operation on $\mathcal{N}(E)$. $\mathcal{N}(E)$ is called the Nambooripad semigroup of E.

Nambooripad's analogue of the fundamental representation (see Remark 1.25) is [51], [52]:

Theorem 12.7. (i) If E is a regular biordered set, then $\mathcal{N}(E)$ is a fundamental regular semigroup with biordered set of idempotents E.

(ii) Let S be a regular semigroup with biordered set of idempotents $E = E(S)$. For $a \in S$ with inverse a^-, let $\theta_S(a) = [aa^-, \gamma, a^-a]$ where $\gamma(x) = a^-xa$. Then $\theta_S: S \to \mathcal{N}(E)$ is a well-defined, idempotent separating homomorphism with kernel μ.

Definition 12.8. Let E be a biordered set. Then a map $*: E \to E$ is an involution if

(i) $(e^*)^* = e$ for all $e \in E$.

(ii) If $e,f \in E$, with ef defined, then f^*e^* is defined and $(ef)^* = f^*e^*$.

Corollary 12.9. Let M be a fundamental unit regular monoid with group of units $G, E = E(M)$. Let $*: E \to E$ be an involution. Then $*$ extends to an involution of M if and only if for all $x \in G$, there exists $y \in G$ such that $(x^{-1}ex)^* = y\, e^*\, y^{-1}$.

Proof. The necessity of the condition being obvious, we prove sufficiency. By general considerations, $*$ extends to an involution of $\mathcal{N}(E)$. Since M is fundamental, we may assume by Theorem 12.7 that M is a submonoid of $\mathcal{N}(E)$. Let $x \in G$. Then there exists $y \in G$ such that $ye^*y^{-1} = (x^{-1}ex)^* = x^*e^*(x^*)^{-1}$ for all $e \in E$. By Remark 1.21 (ii), $x^* = y \in G$. Since $M = EG$, $M^* = M$.

Definition 12.10. A semigroup S is locally inverse if S is regular and eSe is an inverse semigroup for all $e \in E(S)$.

The next result is from Nambooripad [52; Theorem 7.6].

Corollary 12.11. Let E be a biordered set. Then the following conditions are equivalent.

(i) eEe is a semilattice for all $e \in E$.

(ii) $|\text{sand}(e,f)| = 1$ for all $e,f \in E$.

(iii) $E \cong E(S)$ for some locally inverse semigroup S.

Definition 12.12. A local semilattice is a biordered set E such that $|\text{sand}(e,f)| = 1$ for all $e,f \in E$.

Remark 12.13. (i) A related, more general system (which does not always yield a semigroup) was considered by Schein [105]. Local semilattices were first considered by Nambooripad [52]–[55]. See also [46], [47].

(ii) Locally inverse semigroups have been much studied. See for example [45], [60], [119].

The following definition is due to the author [79].

Definition 12.14. Let $\Omega = (\Omega,\leq)$ be a $\wedge$–semilattice with a minimum element 0. Let $\perp$ be a symmetric relation defined on Ω such that $0 \perp 0$. Then $\Omega = (\Omega,\perp)$ is a parabolic semilattice if the following conditions hold.

(i) $\alpha\Omega = \{\beta \in \Omega \mid \beta \leq \alpha\}$ is finite (and hence a lattice) for all $\alpha \in \Omega$.

(ii) If $\gamma, \alpha_1, \alpha_2, \beta_1, \beta_2 \in \Omega$, $\alpha_1 \perp \alpha_2$, $\beta_1 \perp \beta_2$, $\alpha_1 \geq \beta_1$, $\gamma \geq \alpha_2$, $\gamma \geq \beta_2$, then $\alpha_2 \geq \beta_2$.

(iii) If $\alpha_1, \alpha_2, \beta_1 \in \Omega$, $\alpha_1 \perp \alpha_2$, $\alpha_1 \geq \beta_1$, then there exists $\beta_2 \in \Omega$ (unique by (ii)) such that $\alpha_2 \geq \beta_2$, $\beta_1 \perp \beta_2$.

(iv) If $\alpha, \alpha_1, \alpha_2, \beta. \beta_1, \beta_2 \in \Omega$, $\alpha \geq \alpha_i$, $\beta \geq \beta_i$, $\alpha_i \perp \beta_i$, $i = 1,2$, then $(\alpha_1 \vee \alpha_2) \perp (\beta_1 \vee \beta_2)$.

Remark 12.15. Let G be a reductive group, $\Omega = \Omega_G$ the set of all parabolic subgroups of G. If $P_1, P_2 \in \Omega$, define $P_1 \leq P_2$ if $P_2 \subseteq P_1$. Define $P_1 \perp P_2$ if P_1, P_2 are opposite (see Definition 4.40). Then $(\Omega,\perp)$ is a parabolic semilattice.

Definition 12.16. Let $\Omega = (\Omega,\perp)$ be a parabolic semilattice, set $E_\Omega = \{(\alpha,\alpha')\,|\,\alpha,\alpha' \in \Omega,\ \alpha \perp \alpha'\}$. If $e = (\alpha,\alpha')$, $f = (\beta,\beta') \in E_\Omega$, then define $f \leq_r e$ if $\beta \leq \alpha$, $f \leq_\ell e$ if $\beta' \leq \alpha'$. If $f \leq_r e$ then let $ef = f$, $fe = (\beta,\beta^-)$ where $\beta^- \in \Omega$ is such that $\beta \perp \beta^-$, $\beta^- \leq \alpha'$. If $f \leq_\ell e$, then let $fe = f$, $ef = (\beta_1,\beta')$ where $\beta_1 \in \Omega$ is such that $\beta_1 \perp \beta'$ amd $\beta_1 \leq \alpha$.

The following result is due to the author [79].

Theorem 12.17. Let Ω be a parabolic semilattice. Then E_Ω is a local semilattice with an involution.

Proof. The involution on $E = E_\Omega$ is given by the map: $(\alpha,\alpha') \to (\alpha',\alpha)$. Let $e = (\alpha,\alpha')$, $f = (\beta,\beta') \in \Omega$ such that $e \leq_r f \leq_\ell e$. Then $\alpha \leq \beta$, $\beta' \leq \alpha'$. By Definition 12.14 (iii), there exists $\beta^- \in \Omega$ such that $\alpha' \leq \beta^-$, $\beta \perp \beta^-$. Then $\beta' \leq \beta^-$. By 12.14 (ii), $\beta^- = \beta'$. Hence $\alpha' = \beta'$. So again by 12.14 (ii), $\alpha = \beta$. So $e = f$. It now follows easily that the partial binary operation of Definition 12.16 is well-defined. Let $\leq\, = \,\leq_r \cap \leq_\ell$, $\mathcal{R} = \leq_r \cap (\leq_r)^{-1}$, $\mathcal{L} = \leq_\ell \cap (\leq_\ell)^{-1}$. Clearly 12.1 (i), (ii) hold. Let $e = (\alpha,\alpha^-)$, $f = (\beta,\beta^-)$, $g = (\gamma,\gamma^-) \in E$. Suppose first that $f \leq_r e$, $g \leq_r e$, $g \leq_\ell f$. Then $\beta \leq \alpha$, $\gamma \leq \alpha$, $\gamma^- \leq \beta^-$. We see by 12.14 (ii) that $\gamma \leq \beta$. So $g \leq f$. Now if $ge = (\gamma,\gamma')$, $fe = (\beta,\beta')$, then $\beta' \leq \alpha^-$, $\gamma' \leq \alpha^-$. By 12.14 (ii), $\gamma' \leq \beta'$. So $ge \leq fe$. Hence

$$f \leq_r e,\ g \leq_r e,\ g \leq_\ell f \text{ imply } g \leq f,\ ge \leq fe \tag{11}$$

In particular 12.1 (iii) holds. Next assume that $g \leq_r f \leq_r e$. Then $\gamma \leq \beta \leq \alpha$. If $ge = (\gamma,\gamma')$, then $\gamma' \leq \alpha^-$. Let $(ge)f = (\gamma,\gamma'')$. Then $\gamma'' \leq \beta^-$. So by definition, $gf = (\gamma,\gamma'') = (ge)f$. Hence 12.1 (iv) also holds.

Now let $e = (\alpha,\alpha^-)$, $f = (\beta,\beta^-) \in E$. Then $fEe \subseteq \beta\Omega \times \alpha^-\Omega$ is finite. Let $fEe = \{h_1,\ldots,h_k\}$, $h_i = (\gamma_i,\gamma_i^-)$, $i = 1,\ldots,k$. Then $\gamma_i \leq \beta$, $\gamma_i^- \leq \alpha^-$, $i = 1,\ldots,k$. So

$\gamma = \gamma_1 \vee \ldots \vee \gamma_k \perp \gamma^- = \gamma_1^- \vee \ldots \vee \gamma_k^-$ by 12.14 (iv). Clearly $h = (\gamma, \gamma^-) \in fEe$, $h_i \leq h$, $i = 1,\ldots,k$. It follows that $\text{sand}(e,f) = \{h\}$. Now let $g \in E$ and suppose that $e \leq_r g$, $f \leq_r g$. Then $h \leq_r g$, $h \leq_\ell e$. So by (11), $h \leq e$, $hg \leq eg$. Also, $hg \mathcal{R} h \leq_r f \mathcal{R} fg$, whereby $hg \leq_r fg$. Hence $hg \in fg\,E\,eg$. Let $\text{sand}(eg,fg) = \{h'\}$. We claim that $h' = hg$. Now

$$h \mathcal{R} hg \leq h' \leq_r fg \mathcal{R} f \leq_r g$$

So $h' \leq_r g$. Also, $h' \leq_\ell eg \leq g$. So by (11), $h' \leq eg \mathcal{R} e$. Thus $h' \leq_r e$ and $h'e \leq e$. Now $h'e \mathcal{R} h' \leq_r f$. So $h'e \leq_r f$ and $h'e \in fEe$. Hence $h'e \leq h$. So $h' \mathcal{R} h'e \leq h$ and $h' \leq_r h$. Thus $h' \mathcal{R} h \mathcal{R} hg$. Now $hg \leq eg$, $h' \leq eg$. So by the dual of (11), $h' = hg$. Thus 12.1 (v) holds. It follows that E is a local semilattice, proving the theorem.

Definition 12.18. Let G be a reductive group, Ω the parabolic semilattice of all parabolic subgroups of G. Set $E_G = E_\Omega$. We call E_G the local semilattice of G.

Definition 12.19. Let E be a biordered set.

(i) Let $\langle E \rangle$ denote the subsemigroup of $\mathcal{N}(E)$ generated by E.

(ii) Let $\sim$ denote the transitive closure of $\mathcal{R} \circ \mathcal{L}$ on E and let $\mathcal{U}(E) = \mathcal{U}(\langle E \rangle) = E/\sim$. For $e \in E$, let $[e]$ denote the $\sim$-class of e.

Remark 12.20. (i) By Nambooripad [52], $\langle E \rangle$ is the unique fundamental, idempotent generated regular semigroup having E as the biordered set of idempotents. By [21; Lemma 2.2], $\sim$ is just the Green's relation $\mathcal{D}$ on $\langle E \rangle$ restricted to E.

(ii) Let E be a local semilattice. Suppose that $\sim$ restricted to eEe is equality for all $e \in E$. Then it follows from (i) that $e \langle E \rangle e$ is a semilattice for all $e \in E$. See [54].

13 TITS BUILDINGS

The collection of parabolic subgroups of a reductive group has a rich geometric structure. This structure is rather amazingly preserved, via Tits buildings, in a much larger class of groups, including all finite simple groups of Lie type. In this chapter we will briefly describe the theory of Tits buildings [115] and explain how a local semilattice can be naturally associated with such a system.

By a complex is meant a $\wedge$-semilattice $\Omega = (\Omega,\leq)$ with a minimum element 0 such that for all $\alpha \in \Omega$, $\alpha\Omega = \{\beta \in \Omega | \beta \leq \alpha\}$ is a finite Boolean lattice. The minimal elements of $\Omega\backslash\{0\}$ are called vertices. If $\alpha \in \Omega$, then the rank of α is defined to be the number of vertices in $\alpha\Omega$. The maximal elements of Ω are called chambers. We will assume that all chambers are of the same rank d and that every element of Ω is $\leq$ a chamber. We define the rank of Ω to be d. Let α, α' be chambers. We will assume that Ω is connected, i.e. there exist chambers $\alpha = \alpha_0,\alpha_1,\ldots,\alpha_m = \alpha'$ such that $\alpha_i \wedge \alpha_{i+1}$ has rank $d-1$ for $i = 0,\ldots,m-1$. If m is minimal, then we set $\mathrm{dist}(\alpha,\alpha') = m$. A non-empty subset Ω' with the property that $\alpha\Omega' \subseteq \Omega'$ for all $\alpha \in \Omega'$ is said to be a subcomplex. Ω is said to be thick if every element of rank $d-1$ is less than at least three chambers. Ω is said to be thin if every element of rank $d-1$ is less than exactly two chambers.

Definition 13.1. A (Tits) building is a pair $\Omega = (\Omega, \mathcal{A})$ where Ω is a complex, $\mathcal{A}$ a family of finite subcomplexes called apartments such that

(i) Ω is thick.

(ii) Each apartment Σ is thin.

(iii) Any two elements of Ω belong to an apartment.

(iv) If $\Sigma, \Sigma' \in \mathcal{A}$ and if $\alpha,\beta \in \Sigma \cap \Sigma'$, then there exists an isomorphism $\phi: \Sigma \to \Sigma'$ such that $\phi(\gamma) = \gamma$ for all $\gamma \in \alpha\Omega \cup \beta\Omega$.

Example 13.2. Let G be a reductive group, Ω the set of all parabolic subgroups of G ordered by reverse inclusion. If T is a maximal torus of G, let $\Sigma(T) = \{P \in \Omega \mid T \subseteq P\}$. Let $\mathcal{A} = \{\Sigma(T) \mid T$ is a maximal torus of $G\}$. Then $(\Omega,\mathcal{A})$ is a building.

We briefly list some basic facts concerning buildings. We refer to [111], [115] for details. Let $\Sigma \in \mathcal{A}$, α a chamber in Σ. Then there exists a unique retraction $\pi_\alpha: \Sigma \to \alpha\Sigma$, i.e. (i) $\pi_\alpha(\beta) = \beta$ for all $\beta \in \alpha\Sigma$, and (ii) π_α restricted to $\alpha'\Sigma$ is an isomorphism for any chamber $\alpha' \in \Sigma$. If $\beta,\beta' \in \Sigma$, then β,β' are said to be of the same type, type (β) = type (β'), if $\pi_\alpha(\beta) = \pi_\alpha(\beta')$. This concept is independent of the choice of the chamber α. If $\alpha \in \Sigma$ is a chamber, then there exist a unique $\alpha' \in \Sigma$ called the opposite of α in Σ such that $\mathrm{dist}(\alpha,\alpha')$ is maximum. There is a unique automorphism $\theta: \Sigma \to \Sigma$ such that for any chamber α of Σ, α and $\theta(\alpha)$ are opposite. We then define β and $\theta(\beta)$ to be opposite for any $\beta \in \Sigma$. Now let $\beta,\beta' \in \Omega$. Then we define β,β' to be of the same type, type (β) = type (β') if they are of the same type in some (hence every) apartment containing them. β,β' are defined to be opposite $(\beta \perp \beta')$ if they are opposite in some (hence every) apartment containing them. β,β' are of opposite type if β is of the same type as an opposite of β'. In the situation of Example 13.2, these definitions agree with those given in Definitions 4.40, 4.52. If $\alpha,\alpha', \beta,\beta' \in \Omega$ and if $\alpha \perp \alpha'$, $\beta \perp \beta'$, then type (α) = type (β) if and only if type (α') = type (β'). If $\alpha,\beta \in \Omega$, then by [115; Proposition 3.30], type (α) = type (β) if and only if there exists $\gamma \in \Omega$ with $\alpha \perp \gamma$, $\beta \perp \gamma$. If Ω is of rank 1, then any two non–zero elements have the same type and any two non–zero unequal elements are opposite.

It is routinely verified that $(\Omega,\perp)$ is a parabolic semilattice (see Definition 12.14). Hence by Definition 12.16, Theorem 12.17, we can construct a local semilattice $E = E_\Omega$. We follow the notation of Definition 12.19. By the preceding discussion on buildings, we see that $\sim = \mathcal{R} \circ \mathcal{L} \circ \mathcal{R} = \mathcal{L} \circ \mathcal{R} \circ \mathcal{L}$ and that for any $e = (\alpha,\alpha')$, $f = (\beta,\beta') \in E$, $e \sim f$ if and only if type (α) = type (β) if and only if type (α') = type (β'). In particular if $e,f_1,f_2 \in E$, $e \geq f_i$, $i = 1,2$ and if $f_1 \sim f_2$, then $f_1 = f_2$. Thus by Remark 12.20 (ii), $e \prec E \succ e$ is a semilattice for all $e \in E$. Also note that $\mathcal{U}(\prec E \succ) \cong E/\sim$ is a finite Boolean lattice. Thus we have shown the following result of the author [79].

<u>Theorem 13.3</u>. (i) $E = E_\Omega$ is a local semilattice.

(ii) $e \prec E \succ e$ is a semilattice for all $e \in E$. In particular $x^2 = x^3$ for all $x \in \prec E \succ$.

(iii) $\sim = \mathcal{R} \circ \mathcal{L} \circ \mathcal{R} = \mathcal{L} \circ \mathcal{R} \circ \mathcal{L}$ on E and $\mathcal{U}(\prec E \succ) \cong E/\sim$ is a finite Boolean lattice.

Let $\alpha \in \Omega$ be a chamber, $\beta \in \Omega$. Then by [115; Section 3.19], there exists a unique chamber $\alpha' \in \Omega$ such that $\alpha' \geq \beta$, $\mathrm{dist}(\alpha,\alpha')$ is minimum. α' is denoted by $\mathrm{proj}_\beta(\alpha)$. Let $\alpha,\alpha' \in \Omega$ be chambers, $\beta,\beta' \in \Omega$ be of rank $d - 1$. Suppose $\alpha > \beta$, $\alpha' > \beta'$, $\beta \perp \beta'$. Then by [115; Proposition 3.29], $\alpha \perp \alpha'$ if and only if $\mathrm{proj}_\beta(\alpha') \neq \alpha$. Let $E = E_\Omega$, E_{max} the set of maximal elements of $(E,\leq)$.

<u>Lemma 13.4</u>. Let $e = (\alpha,\alpha^\neg) \in E_{max}$, $h = (\beta,\beta^\neg) \in E$, e covers h. Then there exists a unique $f^* = f^*(e,h) \in E_{max}$ such that $ef^* = f^*e = h$ in $\prec E \succ$. Moreover $f^* = (\mathrm{proj}_\beta(\alpha^\neg), \mathrm{proj}_{\beta^\neg}(\alpha))$. Let $f \in E_{max}$, $f \geq h$. Then $ef = h$ if and only if $f\, \mathcal{R}\, f^*$, and $fe = h$ if and only if $f\, \mathcal{L}\, f^*$.

Proof. Since $\alpha \perp \alpha^{-}$, we see that $proj_{\beta^{-}}(\alpha) \neq \alpha^{-}$. But by [115; Theorem 3.28], $\alpha^{-} = proj_{\beta^{-}} proj_{\beta}(\alpha^{-})$. Hence $proj_{\beta}(\alpha^{-}) \perp proj_{\beta^{-}}(\alpha)$. So $f^* = (proj_{\beta}(\alpha^{-}), proj_{\beta^{-}}(\alpha)) \in E_{max}$. Let $f = (\gamma,\gamma^{-}) \in E_{max}$, $f > h$. Then $\gamma > \beta$, $\gamma^{-} > \beta^{-}$. Suppose $f \mathcal{R} f^*$. Then $\gamma \neq proj_{\beta}(\alpha^{-})$. So $\gamma \perp \alpha^{-}$. Hence $e = (\alpha,\alpha^{-}) \mathcal{L} (\gamma,\alpha^{-}) \mathcal{R} (\gamma,\gamma^{-}) = f$. Hence $ef \mathcal{J} e$ in $\langle E \rangle$. Thus $ef \neq h$. Next suppose that $f \mathcal{R} f^*$. Then $\gamma = proj_{\beta}(\alpha^{-})$. So γ is not opposite to α^{-}. It follows that there is no $e_1 \in E$ with $e \mathcal{L} e_1 \mathcal{R} f$. So $ef \not\mathcal{J} e$. Now $efh = hef = h$. Since J_e covers J_f, $ef \mathcal{J} h$. It follows that $ef = h$. This proves the lemma.

If $e,f \in E_{max}$, then define $e \,\delta\, f$ if in $\langle E \rangle$, $ef = fe$ is covered by e. Let δ^* denote the transitive closure of δ. Let $p(e) = \{h \in E \mid h \leq f,\ f \,\delta^* e$ for some $f \in E_{max}\}$. Now let $e = (\alpha,\alpha^{-}) \in E_{max}$. Let Σ be the unique apartment of Ω containing α,α^{-}. If $\beta \in \Omega$, then let β^{-} denote the unique opposite of β in Σ. Let $\hat{\Sigma} = \{(\beta,\beta^{-}) \mid \beta \in \Sigma\}$. By Lemma 13.4, $p(e) = \hat{\Sigma}$. So $(p(e),\leq) \cong \Sigma$. Let λ: $E \to E/\mathcal{R}$ denote the natural map. Let $\mathcal{A}' = \{\lambda(p(e)) \mid e \in E_{max}\}$. We have shown:

Theorem 13.5. $(E/\mathcal{R}, \mathcal{A}')$ is a building isomorphic to $(\Omega, \mathcal{A})$.

Definition 13.6. Let $E = E_{\Omega}$.

(i) Aut* E is the group of all automorphisms σ of E such that $e \sim e\sigma$ for all $e \in E$.

(ii) Aut* $\langle E \rangle$ is the group of all automorphisms σ of the semigroup $\langle E \rangle$ such that $a \mathcal{J} a\sigma$ for all $a \in \langle E \rangle$.

(iii) Aut* Ω is the group of all automorphisms σ of Ω such that type (α) = type $(\alpha\sigma)$ for all $\alpha \in \Omega$.

The following result is due to the author [79].

Theorem 13.7. Aut* $E_{\Omega} \cong$ Aut* $\langle E_{\Omega} \rangle \cong$ Aut* Ω.

Proof. That $Aut^*E_\Omega \cong Aut^* \langle E_\Omega \rangle$ follows from Theorem 12.7. So we need to show that $Aut^*E \cong Aut^*\Omega$. First let $\sigma \in Aut^*\Omega$. Then $\bar{\sigma} \in Aut^*E$ where $(\alpha,\alpha')\bar{\sigma} = (\alpha\sigma,\alpha'\sigma)$. Conversely let $\theta \in Aut^*E$. Then for all $e,e' \in E$, $e \mathcal{R} e'$ if and only if $e\theta \mathcal{R} e'\theta$, and $e \mathcal{L} e'$ if and only if $e\theta \mathcal{L} e'\theta$. It follows that there exist $\sigma_1,\sigma_2 \in Aut^*\ \Omega$ such that $(\alpha,\beta)\theta = (\alpha\sigma_1,\beta\sigma_2)$ for all $(\alpha,\beta) \in E$. We claim that $\sigma_1 = \sigma_2$. For suppose $\alpha\sigma_1 \neq \alpha\sigma_2$ for some $\alpha \in \Omega$. Then $\alpha\sigma_1\sigma_2^{-1} \neq \alpha$. Now α, $\alpha\sigma_1\sigma_2^{-1} \in \Sigma$ for some apartment Σ. So $\alpha \perp \beta$ for a unique $\beta \in \Sigma$. Then $\alpha\sigma_1\sigma_2^{-1}$ is not opposite to β. So $\alpha\sigma_1$ is not opposite to $\beta\sigma_2$. Thus $(\alpha,\beta) \in E$, $(\alpha\sigma_1,\beta\sigma_2) \notin E$. But $(\alpha\sigma_1,\beta\sigma_2) = (\alpha,\beta)\theta \in E$, a contradiction. Hence $\sigma_1 = \sigma_2$. It follows that $Aut^*\ E \cong Aut^*\ \Omega$, proving the theorem.

Definition 13.8. A Tits system is a quadruple (G,B,N,S) where G is a group, B,N are subgroups of G generating G, $T = B \cap N \triangleleft N$, S is a generating set of order 2 elements of $W = N/T$ such that

(i) $\theta B\theta \neq B$ for any $\theta \in S$.

(ii) $\sigma B\theta \subseteq B\sigma B \cup B\sigma\theta B$ for all $\sigma \in W, \theta \in S$.

Remark 13.9. Let G be a reductive group, T a maximal torus of G, B a Borel subgroup of G, $N = N_G(T)$, $\mathcal{S}$ the set of simple reflections relative to B. Then $(G,B,N,\mathcal{S})$ is a Tits system. Moreover any finite simple group of Lie type admits a Tits system. See [111], [115].

Let (G,B,N,S) be a Tits system and assume that the Weyl group $W = N/T$ is finite. If $I \subseteq S$, let $W_I = \langle I \rangle$, $P_I = BW_IB$. Then the P_I's are exactly the subgroups of G containing B. Let $\Sigma = \{\sigma^{-1}P_I\sigma \mid \sigma \in W, I \subseteq S\}$. Let $\mathcal{A} = \{x^{-1}\Sigma x \mid x \in G\}$, $\Omega = \{x^{-1}P_Ix \mid x \in G, I \subseteq S\}$. If $P_1, P_2 \in \Omega$, then define $P_1 \geq P_2$ if $P_1 \subseteq P_2$. Then by [115; Theorem 3.2.6], $\Omega = \Omega_G = (\Omega,\mathcal{A})$ is a building. The elements of Ω are called parabolic subgroups of G. If $P \in \Omega$, then P is conjugate to a unique P_I, $I \subseteq S$. Define type $(P) = I$. Then two parabolic subgroups are of the same type if and only if they are conjugate. The conjugates of B are called <u>Borel subgroups</u>. We call $E_G = E_\Omega$ the <u>local semilattice</u> of G.

<u>Problem 13.10</u>. Let G be a finite simple group of Lie type. Then G acts on E_G by conjugation. So $G \subseteq \mathcal{N}(E_G^1)$. Let M be the submonoid of $\mathcal{N}(E_G^1)$ generated by G and E_G. Then G is the group of units of the fundamental finite regular monoid M and $E(M) = E_G^1$. Study this monoid M. For example, it is always unit regular [85].

14 SYSTEM OF IDEMPOTENTS

Many seemingly unrelated ideas, from the previous chapters, come together in this chapter. Let M be a connected regular monoid with zero 0 and group of units G. Let $E = E(M)$ denote the biordered set of idempotents of M. By Corollary 12.5, E is completely determined by the system $(E, \leq_r, \leq_\ell)$. How much of the structure of M is contained in E ? Let ~, $\mathcal{U}(E)$ be as in Definition 12.19. If $e \in E$, let $[e]$ denote the ~ –class of e. If $e_1, e_2 \in E$, define $[e_1] \geq [e_2]$ if $e_1 \geq e_2'$ for some $e_2' \in [e_2]$. By Theorem 5.9, $\mathcal{U} = \mathcal{U}(E)$ is a lattice, isomorphic to $\mathcal{U}(M)$.

To continue our discussion, we will need the following result from the author [79]. The proof is new.

Lemma 14.1. Let $e,h,f \in E$ such that e covers h covers f. Then there exists a unique $h^* = h^*(e,f) \in E$ such that $e > h^* > f$ and $hh^* = h^*h = f$. Let $h_1 \in E$, $e > h_1 > f$. Then $h_1 \neq h^*$ if and only if there exists $h_2 \in E$, $e > h_2 > f$, such that either $h \mathcal{R} h_2 \mathcal{L} h_1$ or $h \mathcal{L} h_2 \mathcal{R} h_1$. In particular, h^* is determined within E.

Proof. By Theorem 6.7, we are reduced to the case when $e = 1$, $f = 0$. By Remark 8.8, we may assume that G is not a torus. Let $T = C_G(h), E(\overline{T}) = \{1,h,h^*,0\}$, $W = \{1,\sigma\}$, $B = C_G^r(h) = C_G^\ell(h^*)$, $B^- = C_G^\ell(h) = C_G^r(h^*)$. Let $h_1 \in E(M)$, $h_1 \neq 0,1$ such that $h_1h = hh_1 = 0$. By Theorem 11.4,

$$h_1 \in BhB \cup Bh\,\sigma B \cup B\sigma hB \cup Bh^*B$$

Since $Bh = hBh$, $0 \notin h(BhB \cup Bh\sigma B)$, $0 \notin (B\sigma hB)h = B\sigma Bh$. Since $hh_1 = h_1h = 0$, we see that $h_1 \in Bh^*B = Bh^*Bh^*$. So by Theorem 1.1 (i), $h_1 \mathscr{L} h^*$. Similarly, since

$$h_1 \in B^-hB^- \cup B^-h\sigma B^- \cup B^-\sigma hB^- \cup B^-h^*B^-,$$

we see that $h_1 \mathscr{R} h^*$. Hence $h_1 = h^*$. So if $h_1 \in E(M)$, $h_1 \neq 0,1,h^*$, then either $hh_1 \mathscr{J} h$ or $h_1h \mathscr{J} h$. We are done by Theorem 1.1 (vi).

Back to our discussion. Recall that a subset Λ of E is a <u>cross-section lattice</u> if: (i) for all $e \in E$, there exists a unique $e' \in \Lambda$ such that $e \sim e'$, (ii) for all $e,f \in \Lambda$, $[e] \geq [f]$ implies $e \geq f$. See Corollary 9.7. Fix a cross-section lattice Λ of E. Let $B = C_G^r(\Lambda)$, $B^- = C_G^\ell(\Lambda)$, $T = C_G(\Lambda)$. Let $\nu(\Lambda)$ denote the smallest subset of E containing Λ and having the property that if $e,h,f \in \nu(\Lambda)$ with e covering h covering f in E, then $h^* = h^*(e,f)$ (of Lemma 14.1) is in $\nu(\Lambda)$. Then by Corollary 8.10, $\nu(\Lambda) = E(\overline{T})$. Then $\mathscr{E} = \mathscr{E}_\Lambda = (\nu(\Lambda), \leq, \sim)$ is just the $\mathscr{E}$–structure of Chapter 10. The point is that we have determined $\mathscr{E}$ completely from E. By Theorem 10.7, the Weyl group W is recovered from $\mathscr{E}$ as its group of automorphisms. By Corollary 10.21, the set of simple reflections relative to B is recovered as $\mathscr{S} = \mathscr{S}(\Lambda) = \{\sigma \in W \mid \sigma \neq 1,\ \sigma \text{ fixes a chain of length } \operatorname{ht} \mathscr{E}-1 \text{ in } \Lambda\}$. Let $\widetilde{\mathscr{U}}$ be as in Definition 10.22.

<u>Definition 14.2</u>. (i) If $\mathscr{V} \in \widetilde{\mathscr{U}}$, Λ a cross–section lattice of E, then <u>type</u> of $\Lambda_{\mathscr{V}} = \mathscr{V}$. Recall that $\Lambda_{\mathscr{V}} = \{e \in \Lambda \mid [e] \in \mathscr{V}\}$

(ii) $\hat{E} = \{\Lambda_{\mathscr{V}} \mid \Lambda$ is a cross–section lattice of E, $\mathscr{V} \in \widetilde{\mathscr{U}}\}$. If $A,A' \in \hat{E}$, then $A \leq_r A'$ if for all $e \in A$, there exists $e' \in A'$ such that $e \mathscr{R} e'$. Similarly define $A \leq_\ell A'$ if for all $e \in A$, there exists $e' \in A'$ such that $e \mathscr{L} e'$. Let

$\leq = \leq_r \cap \leq_\ell$.

(iii) If $e \in E$, let $\lambda(e) \in \tilde{\mathcal{U}}$ denote the intersection of all $\mathcal{V} \in \tilde{\mathcal{U}}$ with $[e] \in \mathcal{V}$.

The following result is due to the author [79].

Theorem 14.3. (i) $\hat{E}$ is a local semilattice and $\theta: \hat{E} \cong E_G$ where $\theta(A) = (C_G^r(A), C_G^\ell(A))$.

(ii) If $A, A' \in \hat{E}$, then $A \sim A'$ if and only if $\text{type}(A) = \text{type}(A')$. In particular $\mathcal{U}(\hat{E}) \cong \tilde{\mathcal{U}}$.

(iii) If $e \in E$, Λ a cross–section lattice of E containing e, then $C_G^r(e) = C_G^r(\Lambda_{\lambda(e)})$, $C_G^\ell(e) = C_G^\ell(\Lambda_{\lambda(e)})$. If Λ' is any other cross–section lattice of E containing e, then $\Lambda'_{\lambda(e)} = \Lambda_{\lambda(e)}$.

Proof. (i) If $A \in \hat{E}$, then $C_G^r(A) \cap C_G^\ell(A) = C_G(A)$ is a reductive group, whereby $\theta(A) \in E_G$. Now let P, P^- be opposite parabolic subgroups of G relative to a maximal torus T of G. Let $B \subseteq P$ be a Borel subgroup of G containing T, B^- its opposite, relative to T. By Theorem 4.51, there exist $I \subseteq \mathcal{S} = \mathcal{S}(B)$ such that $P = BW_IB$, $P^- = B^-W_IB^-$. There exists, by Theorem 9.10, a cross–section lattice $\Lambda \subseteq E(\overline{T})$ such that $B = C_G^r(\Lambda)$, $B^- = C_G^\ell(\Lambda)$. Let $\mathcal{V} = \{J \in \mathcal{U} \mid e^\sigma = e$ for all $\sigma \in I$, $e \in J \cap \Lambda\}$. Then by Remark 10.23 (ii), $P = C_G^r(\Lambda_{\mathcal{V}})$, $P^- = C_G^\ell(\Lambda_{\mathcal{V}})$. It follows that θ is surjective.

Next let $A, A' \in \hat{E}$, $\mathcal{V} = \text{type}(A)$, $\mathcal{V}' = \text{type}(A')$. Now $A = \Lambda_{\mathcal{V}}$, $A' = \Lambda'_{\mathcal{V}'}$ for some cross–section lattices Λ, Λ' of A. By Corollary 9.7, $x\Lambda x^{-1} = \Lambda'$ for some $x \in G$. Suppose that $A \leq_r A'$. Then $\mathcal{V} \subseteq \mathcal{V}'$ and $xAx^{-1} \subseteq A'$. So $xex^{-1} \mathcal{R} e$ for all $e \in A$. Hence $xe = exe$ for all $e \in A$. So $x \in C_G^r(A)$ and $C_G^r(A') \subseteq C_G^r(xAx^{-1}) = xC_G^r(A)x^{-1} = C_G^r(A)$. Thus $\theta(A) \leq_r \theta(A')$. Assume conversely that $\theta(A) \leq_r \theta(A')$. So $x\, C_G^r(\Lambda_{\mathcal{V}'})x^{-1} = C_G^r(\Lambda'_{\mathcal{V}'}) = C_G^r(A') \subseteq C_G^r(A) =$

$C^r_G(\Lambda_{\mathcal{V}})$. By Theorem 4.51 (iii), $x \in C^r_G(\Lambda_{\mathcal{V}})$ and $\mathcal{V} \subseteq \mathcal{V}'$. So for all $e \in \Lambda_{\mathcal{V}}$, $e \mathcal{R} xex^{-1} \in \Lambda'_{\mathcal{V}} \subseteq \Lambda'_{\mathcal{V}'}$. Hence $A = \Lambda_{\mathcal{V}} \leq_r \Lambda'_{\mathcal{V}'} = A'$. Similarly, $A \leq_\ell A'$ if and only if $\theta(A) \leq_\ell \theta(A')$. In particular $\theta(A) = \theta(A')$ implies $A = A'$. Hence θ is an isomorphism.

(ii) Let $A, A' \in \hat{E}$, type $(A) = \text{type}(A') = \mathcal{V}$. Then $A = \Lambda_{\mathcal{V}}$, $A' = \Lambda'_{\mathcal{V}}$ for some cross–section lattices Λ, Λ' of E. By Corollary 9.7, $x^{-1}\Lambda x = \Lambda'$ for some $x \in G$. So $x^{-1}Ax = A'$ and $x^{-1}C^r_G(A)x = C^r_G(A')$. By Theorem 13.3, $\theta(A) \sim \theta(A')$. So $A \sim A'$ by (i).

(iii) Let $e \in \Lambda$, Λ a cross–section lattice of E. Then clearly $\lambda(e) = \{f \in \Lambda \mid f^\sigma = f$ for all $\sigma \in W$ with $e^\sigma = e\}$ and $\theta(\Lambda_{\lambda(e)}) = (C^r_G(e), C^\ell_G(e))$ by Corollary 10.24. The second statement now follows from (i).

<u>Definition 14.4</u>. (i) Let $\hat{\mathcal{U}} = (\mathcal{U}, \lambda, E_G)$ where $\lambda: \mathcal{U} \to \mathcal{U}(E_G)$ is given by $\lambda(J) = [C^r_G(e), C^\ell_G(e)] \in \mathcal{U}(E_G)$.

(ii) Let $E_{\hat{\mathcal{U}}} = \{(u,h) \mid u \in \mathcal{U}, h \in E_{\hat{\mathcal{U}}}, \lambda(u) = [h]\}$. Let $a_1 = (u_1,h_1)$, $a_2 = (u_2,h_2) \in E_{\hat{\mathcal{U}}}$. Define $a_1 \mathcal{R} a_2$ if $u_1 = u_2$ and $h_1 \mathcal{R} h_2$. Define $a_1 \mathcal{L} a_2$ if $u_1 = u_2$ and $h_1 \mathcal{L} h_2$. Define $a_1 \leq a_2$ if $u_1 \leq u_2$ and there exists $h \in E_G$ such that $h_1 \leq h$, $h_2 \leq h$. Define $\leq_r = \mathcal{R} \circ \leq$, $\leq_\ell = \mathcal{L} \circ \leq$.

<u>Remark 14.5</u>. By Theorem 14.3, the system $\hat{\mathcal{U}}$ is determined for E.

The following result is due to L. Renner and the author [89].

<u>Theorem 14.6</u>. $E_{\hat{\mathcal{U}}}$ uniquely determines a biordered set and $E \cong E_{\hat{\mathcal{U}}}$. The isomorphism is given by: $e \to (J_e, C^r_G(e), C^\ell_G(e))$.

<u>Proof</u>. Let $\psi: E \to E_{\hat{\mathcal{U}}}$ be given by $\psi(e) = (J_e, C^r_G(e), G^\ell_G(e))$. Let $(J,P,P^-) \in E_{\hat{\mathcal{U}}}$. Then there exists $e \in E(J)$ such that $P = C^r_G(e)$. Then $C^\ell_G(e)$ is opposite to P. So there exists $x \in P$ such that $x^{-1}C^\ell_G(e)x = P^-$. So $(J,P,P^-) = \psi(x^{-1}ex)$ and ψ is

surjective. Let $e,f \in E$. By Corollary 6.19, $e \mathscr{R} f$ if and only if $\psi(e) \mathscr{R} \psi(f)$, and $e \mathscr{L} f$, if and only if $\psi(e) \mathscr{L} \psi(f)$. In particular, ψ is bijection. Next assume that $e \geq f$. Then $J_e = J_f$. Let $P = C_G^r(e,f)$, $P^- = C_G^\ell(e,f)$. Then $(P,P^-) \in E_G$, $(P,P^-) \geq (C_G^r(e),C_G^\ell(e))$, $(P,P^-) \geq (C_G^r(f),C_G^\ell(f))$. Hence $\psi(e) \geq \psi(f)$. Finally assume that $\psi(e) \geq \psi(f)$. Then $J_e \geq J_f$ and there exist opposite Borel subgroups B,B^- of G such that $B \subseteq C_G^r(e) \cap C_G^r(f)$, $B^- \subseteq C_G^\ell(e) \cap C_G^\ell(f)$. Now $B = C_G^r(\Lambda)$, $B^- = C_G^\ell(\Lambda)$ for some cross–section lattice Λ of E. Let $T = C_G(\Lambda) = B \cap B^-$. There exist $x \in C_G^r(e)$, $y \in C_G^r(f)$ such that $e_1 = xex^{-1}$, $f_1 = yfy^{-1} \in E(\overline{T})$. Then $e \mathscr{R} e_1$, $f \mathscr{R} f_1$. So by Corollary 6.19, $C_G^r(e) = C_G^r(e_1)$, $C_G^r(f) = C_G^r(f_1)$. By Theorem 9.10, $e_1,f_1 \in \Lambda$. So $\{e_1\} = J_e \cap \Lambda$, $\{f_1\} = J_f \cap \Lambda$. Since $J_e \geq J_f$, $e_1 \geq f_1$. So $f \leq_r e$. Similarly, $f \leq_\ell e$ and $f \leq e$. This completes the proof.

Definition 14.7. Let Aut*E denote the group of all automorphisms σ of E such that $e \sim e\sigma$ for all $e \in E$.

The next result is due to the author [79]. Its proof is simplified due to Theorem 14.6.

Corollary 14.8. $\text{Aut}^*E \cong \text{Aut}^*E_G \cong \text{Aut}^*\Omega_G$.

Proof. If $\sigma \in \text{Aut}^*\Omega_G$, then clearly $(J,P,P^-) \to (J,P\sigma,P^-\sigma)$ is an automorphism of $E_{\hat{\mathscr{U}}} \cong E$. If $\sigma \in \text{Aut}^*E$, then the map: $A \to A\sigma = \{e\sigma \mid e \in A\}$ is an automorphism of $\hat{E} \cong E_G$. The result now follows from Theorem 13.7.

Example 14.9. Let $M = \mathscr{M}_n(K)$. Then $\hat{E}$ is the local semilattice of all chains of idempotents in M containing $0,1$ and $\lambda: \mathscr{U}(M) \to \mathscr{U}(E_G)$ linearly orders the maximal parabolic subgroups of G.

Renner [96] has shown that any normal connected regular monoid with zero admits an involution. See Corollary 16.14. Without the assumption of normality, we now prove a weaker result.

Corollary 14.10. Let M be a connected regular monoid with zero and group of units G. Let T be a maximal torus of G. Then M admits an abstract semigroup involution $*$, such that $t^* = t$ for all $t \in \overline{T}$.

Proof. By Proposition 4.54, G admits an involution $*$ such that $t^* = t$ for all $t \in T$ and so that P^* is opposite to P for any parabolic subgroup P of G containing T. Hence P, P^* are parabolic subgroups of opposite type for any parabolic subgroup P of G. We therefore have an involution $*$ on $E_{\hat{\mathcal{U}}}$ given by $(J,P,P^-)^* = (J,(P^-)^*,P^*)$. By Theorem 14.6, we have an involution $*$ on E. If $e \in E$, then $C_G^r(e^*) = C_G^\ell(e)^*$, $C_G^\ell(e^*) = C_G^r(e)^*$, $e \sim e^*$. It is also clear that $e^* = e$ for all $e \in E(\overline{T})$ and that

$$(x^{-1}ex)^* = x^*e^*(x^*)^{-1} \text{ for all } x \in G, e \in E \tag{12}$$

By Corollary 12.9, $*$ extends to an involution of the monoid M/μ. The problem now is to lift the involution to M.

First let $e,f \in E$ such that $e \mathcal{R} f$. By Theorem 7.1, $e,f \in \overline{B}$ for some Borel subgroup B of G. Let $B_1 = \{x \in B \mid xe = e\}^c$. Then $e,f \in \overline{B}_1$ by Theorem 6.7. Since B_1 is solvable, we see by Corollary 6.8, that there exists a unipotent element $u \in B_1$ such that $eu = f$. Now $u^* \in C_G^r(e)^* = C_G^\ell(e^*)$ is unipotent. So u^*e^* is a unipotent element of the $\mathcal{H}$-class of e^* and $f^*\mu\, u^*e^*$. By Theorem 7.9 (iii), $f^* = u^*e^*$. Hence for $e,f \in E$, $e \mathcal{R} f$, implies that there exists $u \in G$ such that

$$eu = f,\ ue = e,\ f^* = u^*e^*. \tag{13}$$

Now let $e \in E(\overline{T})$. Then $C_G(e)^* = C_G(e)$, $X = \{y \in G \mid ye = ey = e\}$ is a closed normal subgroup of $C_G(e)$. Hence $X^* = X$ by Remark 4.55 (ii). Now let

$x \in G$ such that $xe = e$. Then $f = ex \in E$, $e \mathcal{R} f$. So by (13), $f = eu$ for some $u \in G$ with $ue = e$, $u^*e = f^*$. Then $ux^{-1} \in X$ and hence $(ux^{-1})^* \in X$. Thus $e = (x^*)^{-1}u^*e = (x^*)^{-1}f^*$. But $f = x^{-1}ex$. So by (12), $f^* = x^*e(x^*)^{-1}$. Hence $e = ex^*$. So using (12) we have,

$$xe = e,\ e \in E,\ x \in G \text{ imply } e^*x^* = e^* \tag{14}$$

Now let $e,f \in E$, $x,y \in G$ such that $ex = fy$. So $e \mathcal{R} f$. By (13), there exists $u \in G$ such that $f = eu$, $u^*e^* = f^*$. So $euyx^{-1} = e$, and by the right–left dual of (14), $x^*e^* = y^*f^*$. Let $a \in M$. Then there exists $e \in E$, $x \in G$ such that $a = ex$. Define $a^* = x^*e^*$. By the preceding argument, this is well–defined. By (12), $(xay)^* = y^*a^*x^*$ for all $x,y \in G$, $a \in M$. Let $e \in E(M)$, $x \in G$ such that $exe \mathcal{H} e$. Then $exe = ey$ for some $y \in C_G(e)$. Then $exy^{-1}e = e$, $C_G(e^*) = C_G(e)^*$, $e \mathcal{R} exy^{-1}$. So $e^* \mathcal{L} (exy^{-1})^* = (y^{-1})^*x^*e^*$. Thus $e^* = e^*(y^{-1})^*x^*e^*$ and $(ey)^* = y^*e^* = e^*x^*e^*$. So $(exe)^* = e^*x^*e^*$ if $exe \mathcal{H} e$, $x \in G$. It follows that $(ab)^* = b^*a^*$ for all $a,b \in M$ with $a \mathcal{J} b \mathcal{J} ab$. Now it suffices to show that $(ef)^* = f^*e^*$ for all $e,f \in E$. There exist $h_1,h_2 \in E$ such that $h_1 \mathcal{R} ef \mathcal{L} h_2$. Let J denote the $\mathcal{J}$–class of ef. Now $ef = (h_1e)(fh_2)$, h_1e, $fh_2 \in J$. So by the above

$$(ef)^* = (fh_2)^*(h_1e)^*.$$

Now $h_1 \leq_r e$. So $h_1^* \leq_\ell e^*$ and $(h_1e)^* = e^*h_1^*$ (since $*$ is an involution of E). Similarly $(fh_2)^* = h_2^*f^*$. So

$$(ef)^* = h_2^*f^*e^*h_1^*$$

Since $*$ is an involution of M/μ, $h_1^* \mathcal{L} f^*e^* \mathcal{R} h_2^*$. It follows that $(ef)^* = f^*e^*$, completing the proof.

Conjecture 14.11. Let M be a connected regular monoid with zero. Then there exists a connected regular monoid M' with zero such that $(E(M), (\leq_r)^{-1}, (\leq_\ell)^{-1}) \cong (E(M'), \leq_r, \leq_\ell)$. Examples 8.5, 8.6 form such a pair.

The next result is due to the author [79].

Theorem 14.12. Let M be a connected regular monoid with zero and let $S = M\backslash G$. Then $\mathscr{U}(S)$ is a Boolean lattice if and only if $E(S) \cong E_G$. In such a case, S is a locally inverse semigroup.

Proof. Suppose first that $\mathscr{U}(S)$ is a Boolean lattice. Then by Theorem 9.5, eMe has a solvable group of units for all $e \in E(S)$. Being regular, eMe is commutative. Hence S is a locally inverse semigroup. Moreover, for any $e, f_1, f_2 \in E(S)$, $e \geq f_1$, $e \geq f_2$, $f_1 \mathscr{J} f_2$ imply $f_1 = f_2$. Define $\phi: E(S) \to E_G$ as $\phi(e) = (C_G^r(e), C_G^\ell(e))$. Let $e,f \in E(S)$. Suppose $f \leq_r e$. Then by Theorem 6.16, $e,f \in \overline{C_G^r(e)}$. Let B be any Borel subgroup of $C_G^r(e)$ and let T be a maximal torus of B. By Theorem 9.10, $B = C_G^r(\Lambda)$ for some cross–section lattice $\Lambda \subseteq E(\overline{T})$. There exists $a \in C_G^r(e)$ such that $e' = aea^{-1} \in E(\overline{T})$. Then $e \mathscr{R} e'$. So by Corollary 6.19, $C_G^r(e) = C_G^r(e')$. So $B \subseteq C_G^r(e')$. By Theorem 9.10, $e' \in \Lambda$. Let $f' \in J_f \cap \Lambda$. Then since $J_e \geq J_f$, we have $e' \geq f'$. Since $f \leq_r e'$, we have $f \mathscr{R} fe' \leq e'$. Hence $e' \geq f'$, $f' \mathscr{J} fe'$ in M. Thus $f' = fe$. Hence $f \mathscr{R} f'$. By Corollary 6.19, $B \subseteq C_G^r(f') = C_G^r(f)$. Since B is an arbitrary Borel subgroup of $C_G^r(e)$, we see by Theorem 4.11, that $C_G^r(e) \subseteq C_G^r(f)$. So $\phi(f) \leq_r \phi(e)$.

Assume now that $e,f \in E(S)$, $\phi(f) \leq_r \phi(e)$. Then $C_G^r(e) \subseteq C_G^r(f)$. Let T be a maximal torus of $C_G^r(e)$ with $e \in E(\overline{T})$. Let J denote the maximum element of $\mathscr{U}(S)$, $A = \{h \in J \cap E(\overline{T}) \mid h \geq e\} = \{h_1,\ldots,h_k\}$. Since $E(\overline{T})$ is relatively complemented, $e = h_1 \ldots h_k$. There exists $a \in C_G^r(f)$ such that $f' = afa^{-1} \in E(\overline{T})$. Then $f \mathscr{R} f'$. Let $h \in A$. Then $h \geq e$. By Corollary 9.4, there exists a cross–section lattice $\Lambda \subseteq E(\overline{T})$ such that $e,h \in \Lambda$. So $B = C_G^r(\Lambda) \subseteq C_G^r(e) \subseteq C_G^r(f) =$

$C^r_G(f')$. By Theorem 9.10, $f' \in \Lambda$. Since $J \geq J_{f'}$, we see that $h \geq f'$. So $e = h_1 \ldots h_k \geq f'$. Hence $f \leq_r e$. Similarly, $\phi(f) \leq_\ell \phi(e)$ if and only if $f \leq_\ell e$. In particular, ϕ is injective. Now let $(P,P^-) \in E_G$. Then by Theorem 10.20, there exists a chain Γ in $E(S)$ such that $P = C^r_G(\Gamma)$, $P^- = C^\ell_G(\Gamma)$. Let e denote the maximum element of Γ. Then by the above, $(P,P^-) = (C^r_C(e), C^\ell_G(e)) = \phi(e)$. Hence $E(S) \cong E_G$. Conversely, suppose $E(S) \cong E_G$. Then by Theorem 5.9, $\mathcal{U}(S) \cong \mathcal{U}(E_G)$. But $\mathcal{U}(E_G)$ is a Boolean lattice by Theorem 13.3. This proves the theorem.

<u>Remark 14.13</u>. (i) The monoid in Example 8.6 satisfies the hypothesis of Theorem 14.12.

(ii) Let G_o be a reductive group, $\phi: G_o \to GL(n,K)$ a representation. Let $M(\phi) = \overline{K\phi(G_o)} \subseteq \mathcal{M}_n(K)$. Then $M(\phi)$ is a connected regular monoid with zero. Let $E(\phi) = E(M(\phi))$. Then $E(\phi)$ is a geometrical object, which in light of Theorems 14.3, 14.6, may be viewed as a generalized building. One would conjecture that there exists a representation ϕ of G_o such that $M(\phi)$ satisfies the hypothesis of Theorem 14.12. In such a case $E(\phi)$ is (essentially) the building of G.

Theorems 13.3, 14.6 suggest the following problem: Find all biordered sets naturally constructible from a Tits building. The answer is given by the theory of monoids on groups, developed by the author [85]. We briefly describe some aspects of the theory (without proofs).

Let (G,B,N,S) be a Tits system. A monoid M is a <u>monoid on</u> G if M has G as the group of units and the following three axioms on $E = E(M)$ are satisfied:

(1) If $e,f \in E$, $e \geq f$, then $C^r_G(e,f)$ and $C^\ell_G(e,f)$ are opposite parabolic subgroups of G.

(2) If $e,f \in E$ with $e \mathcal{R} f$ or $e \mathcal{L} f$, then $x^{-1}ex = f$ for some $x \in G$.

(3) $M = \langle E,G \rangle$

For $e,f \in E$, define $e \equiv f$ if $x^{-1}ex = f$ for some $x \in G$. Let

$$\mathcal{U} = \mathcal{U}(M) = E/\equiv$$

For $e,f \in E$, define $[e] \geq [f]$ if $e \geq f_1$ for some $f_1 \in [f]$. The axioms imply that $(\mathcal{U},\leq)$ is a partially ordered set. Consider the type map $\lambda = \lambda(M)\colon \mathcal{U} \to 2^S$ given by $\lambda([e]) = \text{type}(C_G^r(e))$.

Conversely let $\mathcal{U}$ be any partially ordered set with a maximum element 1. Let $\lambda\colon \mathcal{U} \to 2^S$ be any map such that $\lambda(1) = S$. Call λ transitive if for all $J_1,J_2,J_3 \in \mathcal{U}$ with $J_1 \leq J_2 \leq J_3$, we have that any connected component of $\lambda(J_2)$ is either contained in $\lambda(J_1)$ or in $\lambda(J_3)$. Let

$$E(\lambda) = \{(J,P,P^-) \mid J \in \mathcal{U},\ P,P^- \text{ are opposite parabolic subgroups of } G,\ \lambda(J) = \text{type}(P)\}$$

Let $f = (J,P,P^-)$, $e = (J',Q,Q^-) \in E(\lambda)$. Define $f \leq_r e$ if $J \leq J'$ and P is incident to Q (i.e. $P \cap Q$ is parabolic). Define $f \leq_\ell e$ if $J \leq J'$ and P^- is incident to Q^-. First suppose that $f \leq_r e$. Let B_o be a Borel subgroup of G such that $B_o \subseteq P \cap Q$. Let B_o^- be an opposite of B_o contained in P^-. It turns out that there is a unique opposite P' of P (independent of B_o,B_o^-) such that

$$B' = \text{proj}_{Q^-}\text{proj}_Q(B_o^-) \subseteq P'$$

Then define

$$ef = f,\ fe = (J,P,P')$$

If $f \leq_\ell e$, then $fe = f$, ef are defined dually. The following result is due to the author [85].

Theorem 14.14. Let $\mathcal{U}$ be a partially ordered set with a maximum element 1, λ: $\mathcal{U} \to 2^S$ such that $\lambda(1) = S$ and λ is transitive. Then $E(\lambda) \cong E(M)$ for some monoid M on G. Conversely every E(M) is obtained in this manner.

Actually, many of the results from the theory of linear algebraic monoids (for instance, Renner's decomposition) can be generalized to monoids on groups. We refer to [85] for details.

15 $\mathcal{J}$–IRREDUCIBLE AND $\mathcal{J}$–CO–IRREDUCIBLE MONOIDS

In this chapter, we wish to restrict the lattice of $\mathcal{J}$–classes $\mathcal{U}$. Fix a connected regular monoid M with zero 0 and group of units G. The following result is due to Renner [94].

Theorem 15.1. M has a completely reducible, idempotent separating linear representation ϕ such that $\phi(0) = 0$ and for any maximal torus T of G, $\phi|\overline{T}$: $\overline{T} \to \phi(T)$ is an isomorphism.

Proof. By Remark 3.17, we may assume that M is a closed submonoid of some $\mathcal{M}_n(K)$ containing the zero matrix. Let A denote the linear span of M in $\mathcal{M}_n(K)$. Then A is a finite dimensional algebra over K. Let N denote the (nil) radical of A. Then $N \cap M = \{0\}$. Let ψ: $A \to A/N$ denote the natural algebra homomorphism. Let ϕ denote the restriction of ψ to M. Then $\phi^{-1}(0) = \{0\}$. By Theorem 10.12, ϕ is idempotent separating. Let T be a maximal torus of G, T_1 the linear span of T. Then T_1 is a diagonalizable algebra and hence $T_1 \cap N = \{0\}$. So $\psi|T_1$: $T_1 \cong \psi(T_1)$. Since $\overline{T} \subseteq T_1$, $\phi|\overline{T}$: $\overline{T} \cong \phi(\overline{T}) = \overline{\phi(T)}$. Finally, A/N is a finite dimensional semisimple K–algebra and $\phi(M)$ spans A/N. Hence ϕ is completely reducible.

Definition 15.2. M is $\mathcal{J}$–irreducible if $|\mathcal{U}_1(M)| = 1$. M is $\mathcal{J}$–co–irreducible if $|\mathcal{U}^{(1)}(M)| = 1$ (see Definition 6.21). M is $\mathcal{J}$–linear if $\mathcal{U}(M)$ is a chain.

The following result is due to Renner [96; Corollary 8.3.3].

Corollary 15.3. M is $\mathcal{J}$–irreducible if and only if M has an irreducible, idempotent separating linear representation. In such a case $\dim C(G) = 1$.

Proof. Suppose first that M is $\mathcal{J}$–irreducible. Let $\mathcal{U}_1(M) = \{J_o\}$, $T_o = \operatorname{rad} G$. Then by Corollary 6.31, $0 \in \overline{T_o}$. Suppose $\dim T_o > 1$. Then by Theorem 6.20, there exist $e,f \in E(\overline{T_o})$ such that $ef = 0$, $e \neq 0$, $f \neq 0$. If $e_o \in J_o \cap E(\overline{T})$, then by Proposition 6.25, $e \geq e_o$, $f \geq e_o$. So $0 = ef \geq e_o$, a contradiction. Hence $\dim T_o = 1$. Now by Theorem 15.1, M has an idempotent separating, completely reducible linear representation ϕ. Now $\phi = \phi_1 \oplus \ldots \oplus \phi_m$ where each ϕ_i is irreducible. Let $e_o \in E(J_o)$. Then $\phi(e_o) \neq 0$. So $\phi_i(e_o) \neq 0$ for some i. So $\phi_i(e) \neq 0$ for all $e \in E(J_o)$. Let $f \in E(M)$, $f \neq 0$. Then $J_f \geq J_o$. So $f \geq e$ for some $e \in E(J_o)$. Hence $\phi_i(f) \neq 0$. It follows that $\phi_i^{-1}(0) = \{0\}$. By Theorem 10.12, ϕ_i is idempotent separating.

Conversely let $\phi: M \to \operatorname{End}(V)$ be an irreducible, idempotent separating representation of M, where V is a finite dimensional vector space over K. Then $\phi(M)\,(\phi(0)V) = \phi(0)V$ and so $\phi(0) = \{0\}$. Let $J_1, J_2 \in \mathcal{U}_1(M)$, $J_1 \neq J_2$. Let $e_i \in E(J_i)$, $i = 1,2$. Let V_1 denote the span of $\phi(Me_1)V$. Then $V_1 \neq \{0\}$, $\phi(M)V_1 \subseteq V_1$. So $V_1 = V$. But $e_2Me_1 = \{0\}$. So $\phi(e_2)V = \{0\}$, a contradiction.

Remark 15.4. (i) M is $\mathcal{J}$–co–irreducible if and only if $S = M\backslash G$ is a connected semigroup. In such a case, we see as in the proof of Corollary 15.3 that $\dim C(G) = 1$.

(ii) Let $e \in E(M)$, $e \neq 0,1$. If M is $\mathcal{J}$–irreducible, then so is eMe. If M is $\mathcal{J}$–co–irreducible, then by Proposition 6.27, so is M_e.

Definition 15.5. G is nearly semisimple if $\dim C(G) = 1$. G is nearly simple if $\dim C(G) = 1$ and (G,G) is a simple algebraic group (i.e., $G/C(G)$ is a simple group).

Remark 15.6. G is nearly semisimple if and only if $\operatorname{rank}_{ss} G = ht(M) - 1$.

Lemma 15.7. Suppose G is nearly semisimple, $e \in E(M)$, $e \neq 0,1$, H the $\mathcal{H}$–class of e. Then $C_G^r(e)$ is a maximal parabolic subgroup of G if and only if G_e, H are both nearly semisimple.

Proof. By Therorem 6.16, $\operatorname{rank}_{ss} C_G^r(e) = \operatorname{rank}_{ss} C_G(e) = \operatorname{rank}_{ss} H + \operatorname{rank}_{ss} G_e \leq ht(e M e) + ht(M_e) - 2 \leq ht(M) - 2$. The result follows from Theorem 4.51.

Corollary 15.8. Let $e \in E(M)$, $e \neq 0,1$, H the $\mathcal{H}$ –class of e. If M is $\mathcal{J}$–irreducible, then $C_G^r(e)$ is a maximal parabolic subgroup of G if and only if G_e is nearly semisimple. If M is $\mathcal{J}$–co–irreducible, then $C_G^r(e)$ is a maximal parabolic subgroup of G if and only if H is nearly semisimple.

The following result is due to L. Renner and the author [89].

Proposition 15.9. Suppose that M is either $\mathcal{J}$–irreducible or $\mathcal{J}$–co–irreducible. Let $e,f \in E(M)\backslash\{0,1\}$. Then $C_G^r(e) = C_G^r(f)$ if and only if $e \mathcal{R} f$; $C_G^\ell(e) = C_G^\ell(f)$ if and only if $e \mathcal{L} f$.

Proof. We assume that M is $\mathcal{J}$–irreducible, the other case being similar. If $e \mathcal{R} f$, then $C_G^r(e) = C_G^r(f)$ by Corollary 6.19. So assume conversely that $C_G^r(e) = C_G^r(f)$. Let T be a maximal torus of $C_G^r(e)$ such that $e \in E(\overline{T})$. There exists $x \in C_G^r(f)$ such that $f' = xfx^{-1} \in E(\overline{T})$. Then $f \mathcal{R} f'$ and hence $C_G^r(e) = C_G^r(f)$. Let $\mathcal{U}_1(M) = \{J_o\}$, $X = \{h \mid h \in J_o \cap E(\overline{T}), h \leq e\} = \{h_1,\dots,h_k\}$. Since $E(\overline{T})$ is relatively complemented, $e = h_1 \vee \dots \vee h_k$. Let $h \in X$. By Corollary 9.4, there exists $\Lambda \in \mathscr{C}(T)$ with $e,h \in \Lambda$. Then $B = C_G^r(\Lambda) \subseteq C_G^r(e) = C_G^r(f')$. So $f' \in \xi(B) = \Lambda$.

Since $J_{f'} \geq J_{o}$, $f' \geq h$. It follows that $f' \geq e$. Similarly, $e \geq f'$. So $e = f' \mathcal{R} f$.

Corollary 15.10. If M is $\mathcal{J}$-irreducible and $\mathcal{J}$-co-irreducible, then M is $\mathcal{J}$-linear.

Proof. Let Λ be a cross-section lattice of M, $T = C_G(\Lambda)$, $B = C_G^r(\Lambda)$. By Corollary 15.8, Proposition 15.9, $C_G^r(e)$ $(e \in \Lambda \backslash \{0,1\})$ are distinct maximal parabolic subgroups of G containing B. But by Theorem 4.51, G has exactly $\text{rank}_{ss} G$ maximal parabolic subgroups containing B. But by Remark 15.6, $\text{rank}_{ss} G = \text{ht}(M) - 1$. Hence $|\Lambda| - 2 = \text{ht}(M) - 1$ and Λ is linear.

Remark 15.11. Let $n = \text{ht}(M)$. The author [77] has shown that M is $\mathcal{J}$-linear if and only if there exists $a \in M$ such that $a^n = 0$, $a^{n-1} \neq 0$. It is also shown there that in such a case, G is nearly simple of type $A_\ell, B_\ell, C_\ell, F_4$ or G_2.

Lemma 15.12. Suppose G_1, G_2 are closed, connected normal subgroups of G such that $G = G_1G_2$, $(G_1,G_2) = \{1\}$ and $\text{rad}\, G \subseteq G_1$. Let $M_1 = \bar{G}_1$. Then

(i) $E(M_1) = \{e \in E(M) \mid G_2 \subseteq C_G(e)\}$.

(ii) If T_1 is a maximal torus of G_1, $e,f \in E(\bar{T}_1)$, then the join of e,f in $E(\bar{T}_1)$ is the join of e,f in $E(M)$.

Proof. (i) Let $e \in E(M)$ such that $G_2 \subseteq C_G(e)$. Let $x \in C_G(e)$. Then $x = yz$ for some $y \in G_1$, $z \in G_2$. Then $y \in C_G(e)$. So $C_G(e) = C_{G_1}(e)G_2$, $\text{rad}\, G_2 \subseteq \text{rad}\, G \subseteq G_1$. So $\text{rad}\, C_G(e) = \text{rad}\, C_{G_1}(e)\, \text{rad}\, G_2 \subseteq G_1$. By Corollary 6.31, $e \in \bar{G}_1 = M_1$.

(ii) Let h denote the join of e,f in $\overline{C_G(e,f)}$. By Corollary 6.31, it suffices to show that $h \in M_1$. By Proposition 7.5, $G_2 \subseteq C_G(e) \cap C_G(f) \subseteq C_G(h)$. By (i), $h \in M_1$.

Definition 15.13. (i) Let S_1, S_2 be semigroups with zero. Then we denote by $S = S_1 \Delta S_2$ the Rees factor semigroup, $S_1 \times S_2/I$ where $I = S_1 \times \{0\} \cup \{0\} \times S_2$.

(ii) Let E_1, E_2 be biordered sets with identity elements. Then $E = E_1 \nabla E_2$ is the sub–biordered set $(E_1 \backslash \{1\}) \times (E_2 \backslash \{1\}) \cup \{(1,1)\}$ of $E_1 \times E_2$.

Theorem 15.14. Suppose G_1, G_2 are closed connected normal subgroups of G containing rad G such that $G = G_1 G_2$, $(G_1, G_2) = \{1\}$. Let $M_i = \overline{G_i}$, $i = 1,2$. Then

(i) If M is $\mathcal{J}$–co–irreducible, then M_1, M_2 are $\mathcal{J}$–co–irreducible, $E(M) \cong E(M_1) \nabla E(M_2)$ and $\mathcal{U}(M) \cong \mathcal{U}(M_1) \nabla \mathcal{U}(M_2)$.

(ii) If M is $\mathcal{J}$–irreducible, then M_1, M_2 are $\mathcal{J}$–irreducible, $M/\mu \cong M_1/\mu \,\Delta\, M_2/\mu$, $\mathcal{U}(M) \cong \mathcal{U}(M_1) \,\Delta\, \mathcal{U}(M_2)$, and $M = M_1 M_2$.

Proof. (i) Let $J_o = \mathcal{U}^{(1)}(M)$, $E = E(M)$, $E_1 = E(M_1)$, $E_2 = E(M_2)$, $E' = E\backslash\{1\}$, $E_1' = E_1 \backslash \{1\}$, $E_2' = E_2 \backslash \{1\}$. Let $e_i \in E_i'$, $i = 1,2$. Let T_i be a maximal torus of G_i such that $e_i \in E(T_i)$, $i = 1,2$. Since $T_1 T_2$ is a maximal torus of G, $e_1 \vee e_2 \in E$. There exists $h \in E(J_o)$ such that $e_1 \le h$. There exists $e_2' \in E$, $e_2' \mathcal{J} e_2$ such that $e_2' \le h$. There exists $y \in G$ such that $y^{-1} e_2 y = e_2'$. Now $y = y_1 y_2$ for some $y_i \in G_i$, $i = 1,2$. So $e_2' = y_2^{-1} e_2 y_2$, $e_1 = y_2^{-1} e_1 y_2$. So $y_2 h y_2^{-1} \ge e_2$, $y_2 h y_2^{-1} \ge e_1$. Hence $e_1 \vee e_2 \ne 1$ and $e_1 \vee e_2 \in E'$. Now let $e \in E$, $\mathrm{ht}(e) = 1$. Then $\dim eMe = 1$ by Proposition 6.2. By Corollary 15.8, $P = C_G^r(e)$ is a maximal parabolic subgroup of G. By Theorems 4.30, 4.51, either there exists a maximal parabolic subgroup P_1 of G such that $G = P_1 G_2$ or else there exists a maximal parabolic subgroup P_2 of G_2 such that $G = G_1 P_2$. By Theorem 10.20, either there exists $e_1 \in E_1'$ such that $C_G^r(e) = C_G^r(e_1)$ or else there exists $e_2 \in E_2'$ such that $C_G^r(e) = C_G^r(e_2)$. Suppose that there exists $e_1 \in E_1'$ such that $C_G^r(e) = C_G^r(e_1)$. By Proposition 15.9, $e \,\mathcal{R}\, e_1$. Since $G = G_1 G_2$, $(G_1, G_2) = 1$, we see by Corollary 6.8 that $e \in E_1'$. Thus

$$\{e \in E \mid \mathrm{ht}(e) = 1\} \subseteq E_1' \cup E_2'.$$

Since $E(\overline{T})$ is relatively complemented for any maximal torus T of G, we see by Lemma 15.12, that

$$E' = \{e_1 \vee e_2 \mid e_i \in E_i', i = 1,2\}$$

Now let $e_i, f_i \in E_i'$, $i = 1,2$, $e = e_1 \vee e_2$, $f = f_1 \vee f_2$. By Proposition 7.6, there exist maximal tori T_i of G_i, $e_i', f_i' \in E(\overline{T}_i)$ such that $e_i \mathcal{R} e_i'$, $f_i \mathcal{R} f_i'$, $i = 1,2$. There exists $x \in C^r_{G_1}(e_1)$ such that $x^{-1}e_1x = e_1'$. Since T_1T_2 is a maximal torus of G, we see by Proposition 7.5 that $x \in C^r_G(e_1) \cap C_G(e_2) \subseteq C^r_G(e_1 \vee e_2)$. So $e_1' \vee e_2 = x^{-1}(e_1 \vee e_2)x \mathcal{R} e_1 \vee e_2$. Similarly $e_1' \vee e_2 \mathcal{R} e_1' \vee e_2'$. So $e_1 \vee e_2 \mathcal{R} e_1' \vee e_2'$, $f_1 \vee f_2 \mathcal{R} f_1' \vee f_2'$. So if $e_i \mathcal{R} f_i$, $i = 1,2$, then $e_1' = f_1'$, $e_2' = f_2'$ and $e \mathcal{R} f$. Conversely suppose $e \mathcal{R} f$. Then $e_1' \vee e_2' = f_1' \vee f_2' \in E(\overline{T})$ where $T = T_1T_2$. Suppose $e_1' \ngeq f_1'$. Then since $E(\overline{T}_1)$ is relatively complemented, we can find $h \in E(\overline{T}_1)$ such that $1 > h \geq e_1'$, $h \vee f_1' = 1$. So by Lemma 15.12, $h \vee e_2' = h \vee f_1' \vee f_2' = 1$, a contradiction. So $e_1' \geq f_1'$. Similarly $f_1' \geq e_1'$ and $e_1' = f_1'$. For the same reason $e_2' = f_2'$. Hence $e_1 \mathcal{R} f_1$, $e_2 \mathcal{R} f_2$. Similarly $e \mathcal{L} f$ if and only if $e_1 \mathcal{L} f_1$, $e_2 \mathcal{L} f_2$. In particular, $e = f$ implies $e_1 = f_1$, $e_2 = f_2$.

Next suppose $e,f \in E'$ such that $e \geq f$. Then $e,f \in \overline{T}$ for some maximal torus T of G. It follows from the above that there exist $e_i, f_i \in E(\overline{T}) \cap M_i$, $i = 1,2$ such that $e = e_1 \vee e_2$, $f = f_1 \vee f_2$. Suppose $e_1 \ngeq f_1$. Then there exists $h \in E(\overline{T}_1)$ such that $1 > h \geq e_1$, $h \vee f_1 = 1$. Then $1 = h \vee f \leq h \vee e = h \vee e_2$, a contradiction. Hence $e_1 \geq f_1$. Similarly $e_2 \geq f_2$. If conversely, $e_i, f_i \in E_i'$ with $e_i \geq f_i$, $i = 1,2$, then clearly $e_1 \vee e_2 \geq f_1 \vee f_2$. It now follows from Corollary 12.5 that $E \cong E_1 \nabla E_2$. In particular, M_1, M_2 are $\mathcal{J}$-co-irreducible and $\mathcal{U}(M) \cong \mathcal{U}(M_1) \nabla \mathcal{U}(M_2)$.

(ii) Let $E = E(M)$, $E_i = E_i(M)$, $E^* = E\backslash\{0\}$, $E_i^* = E_i\backslash\{0\}$, $M^* = M\backslash\{0\}$, $M_i^* = M_i\backslash\{0\}$, $i = 1,2$. By the proof of (i), $E^* = E_1^*E_2^*$ and for any $e_i, f_i \in E_i^*$, $i = 1,2$, $e_1e_2 \mathcal{R} f_1f_2$ implies $e_i \mathcal{R} f_i$, $i = 1,2$; $e_1e_2 \mathcal{L} f_1f_2$ implies $e_i \mathcal{L} f_i$, $i = 1,2$. In particular $M^* = M_1^* M_2^*$. Now let $a_i, b_i \in M_i^*$, $i = 1,2$ such that $a_1 a_2 \ \mu \ b_1 b_2$. Now $a_i \ \mathcal{R} \ e_i$, $b_i \ \mathcal{R} \ f_i$ in M_i for some $e_i, f_i \in E_i^*$, $i = 1,2$. Then $a_1 a_2 \ \mathcal{R} \ e_1 e_2$, $b_1 b_2 \ \mathcal{R} \ f_1 f_2$. So by the above, $e_1 \ \mathcal{R} \ f_1$, $e_2 \ \mathcal{R} \ f_2$. Hence $a_i \ \mathcal{R} \ b_i$ in M_i, $i = 1,2$. Similarly $a_i \ \mathcal{L} \ b_i$ in M_i, $i = 1,2$. So $a_i \ \mathcal{H} \ b_i$ in M_i, $i = 1,2$. Let $x,y \in M_1$. Then $xa_1ya_2 = xa_1a_2y \ \mu \ xb_1b_2y = xb_1yb_2$. So $xa_1y = 0$ if and only if $xb_1y = 0$. If $xa_1y \neq 0$, then by the above, $xa_1y \ \mathcal{H} \ xb_1y$. Hence $a_1 \ \mu \ b_1$ in M_1. Similarly $a_2 \ \mu \ b_2$ in M_2. It follows that $M/\mu \cong M_1/\mu \ \Delta \ M_2/\mu$. In particular M_1, M_2 are $\mathcal{J}$–irreducible and $\mathcal{U}(M) \cong \mathcal{U}(M_1) \ \Delta \ \mathcal{U}(M_2)$.

Example 15.15. Let $M = \{A \oplus B \mid A,B \in \mathcal{M}_2(K), \det A = \det B\}$, G the group of units of M. Let $G_1 = \{A \oplus B \in G \mid A \text{ is diagonal}\}$, $G_2 = \{A \oplus B \mid B \text{ is diagonal}\}$, $M_i = \overline{G}_i$, $i = 1,2$. Then M is $\mathcal{J}$–co–irreducible but $M \neq M_1 M_2$.

Remark 15.16. Suppose M is $\mathcal{J}$–irreducible, $e \in E(M)$, H the $\mathcal{H}$–class of e. L. Renner has recently shown that $C(H) = e\,C(G)$. It follows from Theorem 7.9 that for $a,b \in M$, $a \ \mu \ b$ if and only if $b = \alpha a$ for some $\alpha \in C(G)$.

Definition 15.17. Let $k \in \mathbb{Z}^+$, $k \leq \text{ht}(M)$. Then M is k–fold $\mathcal{J}$–irreducible if $|\mathcal{U}_i(M)| = 1$ for $i = 1,\dots,k$. M is k–fold $\mathcal{J}$–co–irreducible if $|\mathcal{U}^{(i)}(M)| = 1$ for $i = 1,\dots,k$.

By Theorems 4.30, 15.14, we have,

Corollary 15.18. Suppose that M is either 2–fold $\mathcal{J}$–irreducible or 2–fold $\mathcal{J}$–co–irreducible. Then G is nearly simple.

Corollary 15.19. Suppose that M is 3–fold $\mathcal{J}$–co–irreducible. Then $|\mathcal{U}_1(M)| \leq 2$. Moreover,

(i) M is $\mathscr{J}$–linear if and only if G is of type $A_\ell, B_\ell, C_\ell, F_4$ or G_2.

(ii) If M is 4–fold $\mathscr{J}$–co–irreducible, then G is of type D_ℓ if and only if $|\mathscr{U}_1| = 2$, $|\mathscr{U}_i| = 1$ for $i > 1$.

Proof. Let $\mathscr{U}^{(1)}(M) = \{J_o\}$, Λ a cross–section lattice of M, $T = C_G(\Lambda)$, $B = C_G^r(\Lambda)$, $\mathscr{S} = \mathscr{S}(B)$, $\Gamma = \{e \in \Lambda \mid C_G^r(e)$ is a maximal parabolic subgroup of $G\}$. If $e \in \Gamma$, let $W(C_G(e)) = \{1,\alpha(e)\}$. Then $\alpha\colon \Gamma \to \mathscr{S}$ is a bijection by Theorem 10.20 and Proposition 15.9. If $e \in \Gamma$, then by Lemma 10.16, $\mathscr{S}\backslash\{\alpha(e)\}$ is $\mathscr{S}(C_B(e))$ in $C_G(e)$. Let $\mathscr{S}_o = \{\sigma \in \mathscr{S} \mid \mathscr{S}\backslash\{\sigma\}$ is irreducible$\}$, $\Gamma_o = \{e \in \Gamma \mid (C_G(e), C_G(e))$ is simple$\}$. Then $\alpha(\Gamma_o) = \mathscr{S}_o$. Let $\Lambda \cap J_o = \{e_o\}$. Then e_oMe_o is 2–fold $\mathscr{J}$–co–irreducible and hence by Corollary 15.18, has a nearly simple group of units. Let $\Gamma'_o = \Gamma_o\backslash\{e_o\}$. Then $\Gamma'_o = \{e \in \Lambda \mid ht(e) = 1\}$. Now by the Dynkin diagrams, $|\mathscr{S}_o| = 2$ or 3 with $|\mathscr{S}_o|$ being 2 exactly when G is of type $A_\ell, B_\ell, C_\ell, F_4$ or G_2. Thus $|\Gamma'_o| = 1$ or 2. If $|\Gamma'_o| = 1$, then $\mathscr{U}$ is linear by Corollary 15.10. This proves (i). Now suppose $|\mathscr{S}_o| = 3$ and that M is 4–fold $\mathscr{J}$–co–irreducible. Then $ht(M) > 5$ and G is of type D_ℓ if and only if $\mathscr{S}_o\backslash\{\sigma\}$ is of type $A_{\ell-1}$ for exactly two elements σ of $\mathscr{S}_o$. Now e_oMe_o is 3–fold $\mathscr{J}$–irreducible and hence is not $\mathscr{J}$–linear by (i). So if G is of type D_ℓ, then G_e is of type $A_{\ell-1}$ for $e \in \Gamma_o$. Hence by (i), M_e is $\mathscr{J}$–linear for $e \in \Gamma'_o$. Since $\Gamma'_o = \{e \in \Lambda \mid ht(e) = 1\}$ and $|\Gamma'_o| = 2$, we are done by Corollary 8.11.

Remark 15.20. If M is $\mathscr{J}$–irreducible, the possible $\mathscr{U}(M)$ are determined in [89].

The following result is due to the author [77; Theorem 2.7].

Theorem 15.21. Suppose that M is either $\mathscr{J}$–co–irreducible or 2–fold $\mathscr{J}$–irreducible. Let $S = M\backslash G$, $\dim M = n$. Then every element in S is a product of $2n + 6$

idempotents in S.

Proof. Let $a \in S$. Then $a = e'x$ for some $x \in G$, $e' \in E(S)$. Now $e \geq e'$ for some $e \in E(S)$ with 1 covering e. Then $a = e'ex$. Let H denote the $\mathscr{H}$-class of e. By Theorem 5.9, there exist $e_1, f_1 \in E(S)$ such that $e \mathscr{R} e_1 \mathscr{L} f_1 \mathscr{R} x^{-1}ex$. By Theorem 1.4, $e_1x^{-1}ex \mathscr{H} ex$. So by Remark 1.3 (viii), $ex \in H e_1x^{-1}ex$ and $a \in e'He_1x^{-1}ex$. Thus it suffices to show that every element of H is a product of $2n + 3$ idempotents in S.

So let $e \in E(S)$, 1 covering e, H the $\mathscr{H}$-class of e. Let J_o denote the $\mathscr{J}$-class of e, $E_o = E(J_o)$. By Proposition 5.8, E_o is a closed irreducible subset of M. We have the product maps: $E_o \times \ldots \times E_o \to E_o^k$. Thus $\overline{E_o^k}$ is irreducible for all $k \in \mathbb{Z}^+$. Clearly

$$E_o \subseteq \overline{E_o^2} \subseteq \overline{E_o^3} \subseteq \ldots$$

So there exists $m \in \mathbb{Z}^+$, $m \leq n$ such that $\overline{E_o^m} = \overline{E_o^\ell}$ for all $\ell \in \mathbb{Z}^+$, $\ell \geq m$. Let $S_1 = \overline{E_o^m}$. Then S_1 is a connected semigroup, $e\, S_1\, e$ is a connected monoid. Let $H_1 = H \cap e\, S_1\, e$ denote the group of units of $e\, S_1\, e$. If $x \in C_G(e)$, then $x^{-1}e\, S_1\, ex \subseteq e\, S_1\, e$. So $x^{-1} H_1 x \subseteq H_1$. Since $H = eC_G(e)$ by Theorem 6.16, we see that H_1 is a closed connected normal subgroup of H. Let $X = E_o \times \ldots \times E_o$, the m-fold direct product. Define $\phi: X \to eS_1e$ as $\phi(e_1, \ldots, e_m) = ee_1 \ldots e_m e$. Thus $\phi(X) = eE_o^m e$. So $\overline{\phi(X)} = eS_1e$. By Theorem 2.19, $\phi(X)$ contains a non-empty open set U of eS_1e. Now H_1 is an open subset of eS_1e by Remark 3.21. So $U_1 = U \cap H_1$ is a non-empty open subset of H_1. By Proposition 4.2, $H_1 = U_1^2 \subseteq U^2 \subseteq E_o^{2m+3}$. Thus it suffices to show that $H_1 = H$.

First assume that M is $\mathcal{J}$–co–irreducible. Then by Theorem 6.20, $E(S) = E(S_1)$. Hence $E(eSe) = E(eS_1e)$. So again by Theorem 6.20, the maximal tori of H, H_1 have the same dimension. So by Corollary 4.34, $H = H_1$.

Finally assume that M is not $\mathcal{J}$–co–irreducible. Then $ht(M) > 2$ and M is 2–fold $\mathcal{J}$–irreducible. Let $\mathcal{U}_1(M) = \{J_1\}$. There exists $h \in E(J_1)$ such that $e > h$. By Proposition 6.27, M_h is $\mathcal{J}$–irreducible. So by Corollary 15.3, G_h is nearly semisimple. By Corollary 6.31, the width of e in M_h is greater than 1. So there exists $f \in E(J_o)$ such that $ef = fe, e \neq f, f > h$. Hence $ef \notin J_o$, $ef \neq 0$. Now $ef \in eS_1e = \overline{H}_1$. So $\dim H_1 > 1, H_1 \neq (H_1,H_1)$. Now eMe is 2–fold $\mathcal{J}$–irreducible and hence by Corollary 15.18, H is nearly simple. So by Corollary 4.34, $H = H_1$, proving the theorem.

Remark 15.22. That non–invertible $n \times n$ matrices over a field can be expressed as products of idempotent matrices was first noted by Erdos [22]. See also [13].

Example 15.23. Let $M = \{A \otimes B \mid A,B \in \mathcal{M}_2(K)\}$. Then M is $\mathcal{J}$–irreducible but $\begin{bmatrix} 0 & 1 \\ 1 & 0 \end{bmatrix} \otimes \begin{bmatrix} 1 & 0 \\ 0 & 0 \end{bmatrix}$ is not expressible as a product of idempotents of M.

Remark 15.24. Recently the author [87] has shown that if M is $\mathcal{J}$–irreducible, then the non–units of M are products of idempotents if and only if the group of units G is nearly simple.

16 RENNER'S EXTENSION PRINCIPLE AND CLASSIFICATION

In this chapter, we will need some additional algebraic geometry.

Definition 16.1. Let X,Y be irreducible affine varieties, $\phi: X \to Y$ a dominant morphism. Then

(i) ϕ is finite if K[X] is an integral extension of $\phi^*(K[Y]) \equiv K[Y]$.

(ii) ϕ is birational if there exists a non-empty open subset U of Y such that $\phi|\phi^{-1}(U) : \phi^{-1}(U) \to U$ is an isomorphism.

Remark 16.2. (i) If ϕ is finite, then it is closed (and hence surjective) and the inverse image of any point in Y is a finite set. See [34; Proposition 4.2].

(ii) Let M,M' be connected regular monoids with zero, $\phi: M \to M'$ a dominant homomorphism. If ϕ is finite, the clearly $\phi^{-1}(0) = \{0\}$. The converse has been shown by Renner [91; Proposition 3.4.13]. Equivalently ϕ is idempotent separating (Theorem 10.12). The homomorphism of Example 3.12 is not finite.

Definition 16.3. Let X be an irreducible affine variety. Then

(i) X is normal if K[X] is integrally closed, i.e., the integral closure of K[X] in K(X) is K[X].

(ii) The normalization of X is $\tilde{X} = (\tilde{X},\phi)$ where $\tilde{X}$ is an irreducible normal affine variety, $\phi: \tilde{X} \to X$ is a finite birational morphism.

Remark 16.4. Normality is a local property: X is normal if and only if for all $x \in X$, the ring $\mathcal{O}_x = \{f/g \mid f,g \in K[X], g(x) \neq 0\}$ is integrally closed. In particular if X is normal, then any affine open subset of X is also normal. When X is not normal, the integral closure of $K[X]$ in $K(X)$ is a finitely generated algebra, giving rise to the normalization $\tilde{X}$ of X. See [106; Chapter II, Section 5.2] for details.

Proposition 16.5. Let X be an irreducible affine variety. Then X admits a unique normalization $\tilde{X} = (\tilde{X},\phi)$. Moreover, any dominant morphism from a normal variety into X factors through ϕ uniquely.

The following result is due to Renner [91].

Corollary 16.6. Let M be a connected regular monoid with zero. Then the normalization $\tilde{M} = (\tilde{M},\phi)$ is also a connected regular monoid with zero and ϕ is a finite birational homomorphism.

Proof. We have the product map $p: M \times M \to M$. Hence $p \circ (\phi \times \phi) : \tilde{M} \times \tilde{M} \to M$. By Proposition 16.5, we have the product map $\tilde{p}: \tilde{M} \times \tilde{M} \to \tilde{M}$. Using Proposition 16.5, again, we see that $\tilde{p}$ is associative and that ϕ is a homomorphism. Now $\phi^{-1}(1)$ is a finite group and hence $\phi^{-1}(1)^c = \{1\}$. Thus the group of units of $\tilde{M}$ is reductive. By Theorem 7.4, $\tilde{M}$ is regular. Now $\phi^{-1}(0)$ is a finite ideal of $\tilde{M}$ and hence the zero of $\tilde{M}$.

Remark 16.7. (i) Let $\overline{T}$ be a connected diagonal monoid with zero. Then $\overline{T}$ is normal if and only if, for all $\chi \in \mathcal{X}(T)$, $n \in \mathbb{Z}^+$, $n\chi \in \mathcal{X}(\overline{T})$ implies $\chi \in \mathcal{X}(\overline{T})$. See [37]. In the multiplicative notation this means that the commutative semigroup $\mathcal{X}(\overline{T})$ has the following property: if $a,b \in \mathcal{X}(\overline{T})$, $i \in \mathbb{Z}^+$, then $a^i \mid b^i$ implies $a \mid b$.

(ii) $\overline{T} = \{\text{diag}\ (a,b,c) \mid a,b,c \in K,\ a^2b = c^2\}$ is not normal. Its normalization is $\overline{T}_1 = \{\text{diag}\ (a,b,c) \mid a,b,c \in K,\ ab = c\}$ with $\phi: \overline{T}_1 \to \overline{T}_1$ given by

$\phi(a,b,c) = (a,b^2,c)$.

We will need the following consequence of Zariski's Main Theorem (see [108; Exercise 4.28(3)]).

Proposition 16.8. Let $\phi: X \to Y$ be a bijective, birational morphism of irreducible affine varieties such that Y is normal. Then ϕ is an isomorphism.

Definition 16.9. Let M be a connected regular monoid with zero, Λ a cross–section lattice of M. Then

(i) Λ_{max} denotes the set of maximal elements of $\Lambda\backslash\{1\}$.

(ii) If $e \in \Lambda_{max}$, then $M(e) = M(e,\Lambda) = \{a \in M \mid faf \mathscr{H} f$ for all $f \in e\Lambda\} = \{a \in M \mid \det_{e\Lambda}(a) \neq 0\}$ where $\det_{e\Lambda}$ is as in Definition 3.22.

Remark 16.10. Let G be a reductive group, B, B^- opposite Borel subgroups of G relative to a maximal torus T. Let $U = B_u$, $U^- = B_u^-$. Then B^-B is an affine open subset of G and is called the big cell of G. Moreover, the product map from $U^- \times T \times U$ onto B^-B is an isomorphism of varieties. See [34] for more details on Chevalley's big cell. The open sets $M(e)$ are due to Renner [96]. They will be used in Theorem 16.13 below.

The following result is due to Renner [96; Theorem 4.4].

Theorem 16.11. Let M be a connected normal regular monoid with zero and group of units G. Let Λ be a cross–section lattice of M and set $B = C_G^r(\Lambda)$, $B^- = C_G^\ell(\Lambda)$, $T = C_G(\Lambda)$, $U = B_u$, $U^- = B_u^-$. Then for any $e \in \Lambda_{max}$, we have,

(i) $M(e)$ is an affine open subset of M and $M(e) \subseteq G \cup J_e$.

(ii) $\overline{T}(e) = T \cup eT$ is an affine open submonoid of $\overline{T}$.

(iii) The product map, $p: U^- \times \overline{T}(e) \times U \to M(e)$ given by $p(x,t,y) = xty$ is an isomorphism of varieties.

Proof. (i), (ii) being obvious, we prove (iii). Let H denote the $\mathscr{H}$-class of e. By Theorem 6.16, Lemma 10.16, $Be = eC_B(e) = C_H^r(e\Lambda)$, $eB^- = eC_{B^-}(e) = C_H^{\ell}(e\Lambda)$, $Ue = (Be)_u$, $eU^- = (eB^-)_u$. Let $f \in e\Lambda$, $x \in U^-$, $t \in \overline{T}(e)$, $y \in U$. Then $fx = fxf$, $yf = fyf$. So $fxtyf = (fxf)(ft)(fyf)\ \mathscr{H}\ f$. Hence $xty \in M(e)$. Let $a \in M(e)$. If $a \in G$, then by Theorem 6.33, $a \in B^-B = p(U^- \times T \times U)$. Next suppose $a \in J_e$. Then $eae \in M(e) \cap H$. By Theorem 6.33, $eae \in eB^-Be$. So there exist $x \in C_B(e)$, $y \in C_{B^-}(e)$ such that $eae = eyxe$. Let $b = y^{-1}ax^{-1}$. Then $ebe = e$. So $f = eb \in E(M)$, $e\ \mathscr{R}\ f$. By Proposition 9.8, $f \in \overline{B}$. By Proposition 6.1, there exists $x_1 \in B$ such that $ex_1 = f = eb$. Let $c = bx_1^{-1}$. Then $ec = e$. So $ce \in E(M)$, $e\ \mathscr{L}\ ce$. By Proposition 9.8, $ce \in \overline{B^-}$. By Corollary 6.8, there exists $y_1 \in \overline{B^-}$ such that $y_1e = ce$, $ey_1 = e$. Let $d = y_1^{-1}c$. Then $de = e$, $ed = ey_1^{-1}c = ec = e$. Since $d\ \mathscr{J}\ e$, we see by Theorem 1.4 (i) that $d = e$. Hence $a = yy_1ex_1x \in B^-eB \subseteq p(U^- \times \overline{T}(e) \times U)$. It follows that p is surjective.

By Remark 16.10, p restricted to $U^- \times T \times U$ is an isomorphism onto the open set B^-B of M(e). Hence p is birational. We show next that p is injective. So let $u_1, u_1' \in U^-$, $u_2, u_2' \in U$, $t, t' \in T$ such that $u_1etu_2 = u_1'et'u_2'$. Then $u_1^{-1}u_1'e = etu_2(u_2')^{-1}(t')^{-1}$. So $u_1^{-1}u_1'e = eu_1^{-1}u_1'e = eu_1^{-1}u_1' \in eU^-$. Similarly $u_1^{-1}u_1'e \in Be$. Thus $u_1^{-1}u_1'e \in eU^- \cap Be = eU^- \cap eT = \{e\}$. Now $G_e = T_e \subseteq T$. So $\{z \in C_G(e) \mid ez = ze = e\}$, being a closed normal subgroup of $C_G(e)$ is contained in T by Corollary 4.34. In particular, $u_1^{-1}u_1' \in T \cap U^- = \{1\}$. Thus $u_1 = u_1'$. Similarly $u_2 = u_2'$. So $et = et'$. It follows that p is a bijection. By Remark 16.4, M(e) is normal. So by Proposition 16.8, p is an isomorphism. This proves the theorem.

See [29; Section 5, Lemma 1] for the following.

Proposition 16.12. Let X,Y be irreducible affine varieties, X normal. Let U be a non-empty open subset of X such that $\dim(X \backslash U) \le \dim X - 2$. Then any morphism $\phi: U \to Y$ extends to a morphism $\overline{\phi}: X \to Y$.

The extension principle below is due to Renner [96; Corollary 4.5].

Theorem 16.13. Let M be a connected normal regular monoid with zero and group of units G. Let T be a maximal torus of G. Let S be an algebraic semigroup, ϕ_1: $G \to S$, ϕ_2: $\bar{T} \to S$ homomorphisms such that $\phi_1|T = \phi_2|T$. Then there exists a unique homomorphism ϕ: $M \to S$ extending ϕ_1, ϕ_2.

Proof. Since $\bar{G} = M$, it suffices to extend ϕ_1, ϕ_2 to a morphism ϕ: $M \to S$. Let $\Lambda \subseteq E(\bar{T})$ be a cross–section lattice, $B = C_G^r(\Lambda)$, $B^- = C_G^\ell(\Lambda)$, $U = B_u$, $U^- = B_u^-$. By Corollary 3.28, every ideal of M is closed. Hence by Proposition 6.2, the irreducible components of $M\backslash G$ are exactly $MeM(e \in \Lambda_{max})$. Let M' denote the union of $M(e) \cup G$ ($e \in \Lambda_{max}$). Let X be an irreducible component of $M\backslash M'$. Then $X \subseteq MeM\backslash M(e)$ for some $e \in \Lambda_{max}$. Then $\dim X \le \dim(MeM\backslash M(e) < \dim MeM < \dim M$. Hence $\dim(M\backslash M') \le \dim M - 2$. By Proposition 16.12, it suffices to extend ϕ_1, ϕ_2 to M'. Now for $e,f \in \Lambda_{max}$, $e \ne f$, we have $M(e) \cap M(f) \subseteq G$. Hence by Remark 2.17 (iii), it suffices to find, for $e \in \Lambda_{max}$, a morphism θ: $M(e) \to S$ such that $\theta|\bar{T}(e) = \phi_2|\bar{T}(e)$ and $\theta|B^-B = \phi_1|B^-B$. Define γ: $U^- \times T(e) \times U \to S$ as $\gamma(u_1,t,u_2) = \phi_1(u_1)\phi_2(t)\phi_1(u_2)$. Let $\theta = \gamma \circ p^{-1}$, where p: $U^- \times T(e) \times U \cong M(e)$ is as in Theorem 16.11.

The next result due to Renner [96; Theorem 8.2] follows from Proposition 4.54 and Theorem 16.13.

Corollary 16.14. Let M be a connected normal regular monoid with zero, group of units G. Let T be a maximal torus of G. Then M admits an involution $*$ such that $t^* = t$ for all $t \in \bar{T}$.

Renner [96] uses his extension principle to establish the following classification theorem.

Theorem 16.15. Let G be a reductive group with a maximal torus T and Weyl group W. Let $\overline{T}$ be a normal connected diagonal monoid with zero, having T as the group of units. Suppose the action of W on T extends to $\overline{T}$. Then there is a unique connected normal regular monoid M with zero having G as the group of units such that $\overline{T}$ is the closure of T in M.

This provides a discrete geometrical classification since normal torus embeddings $T \hookrightarrow \overline{T}$ can be viewed geometrically [37], [58]. Let $\dim T = n$, so that $\mathcal{X}(T) \cong \mathbb{Z}^n$. Let $C = \mathcal{X}(\overline{T}) \cup \{0\}$. Then $C = P \cap \mathbb{Z}^n$ for some polyhedral cone $P \subseteq \mathbb{R}^n$. To obtain a monoid, P must be W–invariant. We refer to [95], [96] for details, where the classifying invariants are axiomatized into the concept of a polyhedral root system.

Example 16.16. Consider $G = GL(2,K)$. So $\mathcal{X}(T) \cong \mathbb{Z}^2$. If $W = \{1,\sigma\}$, then σ induces the automorphism σ: $(i,j) \to (j,i)$ on $\mathbb{Z}^2$. Also G has an automorphism: $A \to (A^{-1})^t$ which induces the automorphism: $x \to -x$ of $\mathbb{Z}^2$. Thus Renner's theorem says that there is a one to one correspondence between connected normal regular monoids with zero having G as the group of units and symmetric rational cones (i.e. invariant under σ) in the plane $\mathbb{R}^2$, and with a cone and its negative identified. See [95] for details.

How about connected regular monoids without zero? By Theorem 7.4, the closure of the radical of the group of units is a completely regular monoid. So the first step is to classify connected regular monoids with a solvable group of units. So let M be a connected completely regular monoid such that the group of units G is solvable. Let $G = UT$ where U is the unipotent radical of G and T a maximal torus. Let e denote the minimum idempotent of $E(\overline{T})$. Let $G_1 = UT_e$, $M_1 = \overline{G}_1$. Then by Theorem 6.7 and Corollary 6.8, e is the zero of $\overline{T}_e$ and $E(M) = E(M_1)$. The following classification theorem is due to Renner [102].

<u>Theorem 16.17</u>. Let $G = UT$ be a solvable group. Let ψ denote the set of non–zero weights with respect to the action of T on the Lie algebra of U. Let $\overline{T}$ be a normal connected diagonal monoid with zero having T as the group of units, such that $\psi \subseteq \mathscr{X}(\overline{T}) \cup -\mathscr{X}(\overline{T})$. Then there exist a normal connected completely regular monoid M such that $\overline{T}$ is the closure of T in M. Conversely any normal connected completely regular monoid with G as the group of units and $\overline{T}$ having a zero, is obtained in this manner.

<u>Remark 16.18</u>. If G is not nilpotent and $\dim T = 1$, then there are exactly two such monoids. For example, the monoids $M_1 = \left\{\begin{bmatrix} a & b \\ 0 & 1 \end{bmatrix} \middle| a \in K^*, b \in K\right\}$, $M_2 = \left\{\begin{bmatrix} a & 0 \\ b & 1 \end{bmatrix} \middle| a \in K^*, b \in K\right\}$ have isomorphic groups of units.

<u>Problem 16.19</u>. Study the system of idempotents of connected completely regular monoids.

REFERENCES

[1] Azumaya, G. (1954). Strongly π–regular rings. J. Fac. Sci. Hokkaido Univ. 13, 34–9.

[2] Balinski, M. (1961). On the graph structure of convex polyhedra in n–space. Pacific J. Math. 11, 431–34.

[3] Birkhoff, G. (1948). Lattice theory. Amer. Math. Soc. Colloq. Publ. Vol. 25, New York.

[4] Borel, A. (1969). Linear algebraic groups. Benjamin, New York.

[5] Borel, A. and Tits, J. (1965). Groups réductifs. Publ. Math. I.H.E.S. 27, 55–150.

[6] Carter, R.W. (1985). Finite groups of Lie type: conjugacy classes and complex characters. Wiley.

[7] Chevalley, C. (1956–1958). Classifications des groupes de Lie algebriques. Seminaire Ecole Normale Supérieure. Paris.

[8] Clark, W.E. (1965). Remarks on the kernel of a matrix semigroup. Czechoslovak Math. J. 15(90), 305–10.

[9] Clark, W.E. (1965). Affine semigroups over an arbitrary field. Proc. Glasgow Math. Assoc. 7 , 80–92.

[10] Clifford, A.H. (1941). Semigroups admitting relative inverses. Annals of Math. 42 , 1037–49.

[11] Clifford, A.H. and Preston, G.B. (1961). The algebraic theory of semigroups, Math Surveys, 1, No. 7, Providence, R.I.

[12] Coxeter, H.S.M. (1935). The complete enumeration of finite groups of the form $R_i^2 = (R_iR_j)^{k_{ij}} = 1$. London Math. Soc. 10, 21–5.

[13] Dawlings, R.J.H. (1983). On idempotent affine mappings. Proc. Royal Soc. Edinburgh 93A, 345–48.

[14] DeConcini, C. and Procesi, C. (1985). Complete symmetric varieties II, Advanced Studies in Pure Math 6, 481–513.

[15] DeConcini, C. and Procesi, C. (1986). Cohomology of compactifications of algebraic groups. Duke Math. J. 53, 585–94.

[16] Demazure, M. and Gabriel, P. (1980). Introduction to algebraic geometry and algebraic groups. North–Holland, Amsterdam.

[17] Dieudonne, J. (1974). Cours de géométrie algebrique. 2, Presses Universitaires de France, Paris.

[18] Douglas, R. and Renner, L. (1981). Uniqueness of product and coproduct decompositions in rational homotopy theory. Trans. Amer. Math. Soc. 264, 165–180.

[19] Drazin, M.P. (1961). Pseudo–inverses in associative rings and semigroups. Proc. Cambridge Philos. Soc. 57, 245–50.

[20] Easdown, D. (1985). Biordered sets come from semigroups. J. Algebra 96, 581–91.

[21] Eberhart, C., Williams, W., Kinch, L. (1973). Idempotent generated regular semigroups. J. Austral. Math. Soc. 15, 27–34.

[22] Erdos, J.A. (1967). On products of idempotent matrices. Glasgow Math. J. 8, 118–22.

[23] Fitz–Gerald, D.G. (1972). On inverses of products of idempotents in regular semigroups. J. Austral. Math. Soc. 13, 335–37.

[24] Green, J.A. (1951). On the structure of semigroups. Annals of Math. 54, 163–72.

[25] Grigor'ev, D. Ju. (1981). An analogue of the Bruhat decomposition for the closure of the cone of a classical Chevalley group series. Dokl. Akad. Nauk SSSR 257, No. 5, 1040–44.

[26] Grillet, P.A. (1970). The free envelope of a finitely generated commutative semigroup. Trans. Amer. Math. Soc. 149, 665–82.

[27] Grillet, P.A. (1974). The structure of regular semigroups I. A representation. Semigroup Forum 8, 177–83.

[28] Grillet, P.A. (1974). The structure of regular semigroups II. Cross–connections. Semigroup Forum 8, 254–59.

[29] Grosshans, F.D. (1973). Observable groups and Hilbert's Fourteenth problem. Amer. J. Math. 95, 229–53.

[30] Grünbaum, B. (1967). Convex polytopes. Interscience–Wiley, New York.

[31] Hall, T.E. (1973). On regular semigroups. J. Algebra 24, 1–24.

[32] Hochschild, G.P. (1981). Basic theory of algebraic groups and Lie algebras. Springer–Verlag, New York.

[33] Howie, J. (1976). An introduction to semigroup theory. Academic Press, London.

[34] Humphreys, J.E. (1981). Linear algebraic groups. Springer–Verlag, New York.

[35] Jacobson, N. (1968). Structure of rings. AMS, Colloq. Publ. 37, New York.

[36] Janowitz, M.F. (1966). A semigroup approach to lattices. Canadian J. Math. 8, 1212–23.

[37] Kempf, G., Knudsen, F., Mumford, D., Saint–Donat, B. (1973). Toroidal embeddings I. Springer–Verlag, New York.

[38] Kleiman, E.I. (1984). On connected algebraic monoids. Ural Polytech Inst. VINITI Publ., 14 pp.

[39] Krohn, K. and Rhodes, J. (1965). Algebraic theory of machines I. Prime decomposition theorem for finite semigroups and machines. Trans. Amer. Math. Soc. 116, 450–64.

[40] Lallement, G. (1966). Congrences et equivalences de Green sur un demi–group régulier. C.R. Acad. Sci. Paris, Ser. A, 262, 613–16.

[41] Lallement, G. (1967). Demi–groups reguliers. Ann. Mat. Pura Appl. 77, 47–129.

[42] Leech, J. (1975). H–co–extensions of monoids. Mem. A.M.S. 157.

[43] Loganathan, M. (1982). Idempotent–separating extensions of regular semigroups with abelian kernel. J. Austral. Math. Soc., Ser A 32, 104–13.

[44] Lusztig, G. (1976). On finiteness of the number of unipotent classes. Invent. Math. 34, 201–13.

[45] McAlister, D.B. (1983). Rees matrix covers for locally inverse semigroups. Trans. Amer. Math. Soc. 277, 727–38.

[46] Meakin, J. (1983). The free local semilattice on a set. J. Pure and App. Alg. 27, 263–75.

[47] Meakin, J. and Pastijn, F. (1981). The structure of pseudo–semilattices. Algebra Universalis 13, 335–73.

[48] Miller, D.D. and Clifford, A.H. (1956). Regular $\mathscr{D}$–classes in semigroups. Trans. Amer. Math. Soc. 82, 270–80.

[49] Munn, W.D. (1961). Pseudo–inverses in semigroups. Proc. Cambridge Philos. Soc. 57, 247–50.

[50] Munn, W.D. (1970). Fundamental inverse semigroups. Quart. J. Math. Oxford 21, 157–70.

[51] Nambooripad, K.S.S. (1975). Structures of regular semigroups I. Fundamental regular semigroups. Semigroup Forum 9, 354–63.

[52] Nambooripad, K.S.S. (1979). Structure of regular semigroups I. Memoirs Amer. Math. Soc. 224.

[53] Nambooripad, K.S.S. (1981). Pseudo–semilattices and biordered sets I. Simon Stevin 55, 103–10.

[54] Nambooripad, K.S.S. (1982). Pseudo–semilattices and biordered sets II. Simon Stevin 56, 143–60.

[55] Nambooripad, K.S.S. (1982). Pseudo–semilattices and biordered sets III. Simon Stevin 56, 239–56.

[56] Nambooripad, K.S.S. and Pastijn, F. (1981). V–regular semigroups. Proc. Royal Soc. Edinburgh 88A, 275–91.

[57] Nambooripad, K.S.S. and Pastijn, F. (1985). The fundamental representation of a strongly regular Baer semigroup. J. Algebra 92, 283–302.

[58] Oda, T. (1978). Torus embeddings and applications. Tata Press, Bombay.

[59] Okninski, J. (1985). Strongly π–regular matrix semigroups. Proc. Amer. Math. Soc. 93, 215–17.

[60] Pastijn, F. (1982). The structure of pseudo–inverse semigroups. Trans. Amer. Math. Soc. 273, 631–55.

[61] Petrich, M. (1984). Inverse semigroups. Interscience–Wiley, New York.

[62] Putcha, M.S. (1973). Semilattice decompositions of semigroups. Semigroup Forum 6, 12–34.

[63] Putcha, M.S. (1974). Minimal sequences in semigroups. Trans. Amer. Math. Soc. 189, 93–106.

[64] Putcha, M.S. (1980). On linear algebraic semigroups. Trans. Amer. Math. Soc. 259, 457–69.

[65] Putcha, M.S. (1980). On linear algebraic semigroups II. Trans. Amer. Math. Soc. 259, p. 471–91.

[66] Putcha, M.S. (1982). On linear algebraic semigroups III. Internat. J. Math. and Math. Sci. 4, 667–90; Corrigendum 5, 205–7.

[67] Putcha, M.S. (1982). Green's relations on a connected algebraic monoid. Linear and Multilinear Algebra 12, 205–14.

[68] Putcha, M.S. (1982). The group of units of a connected algebraic monoid. Linear and Multilinear Algebra 12, 37–50.

[69] Putcha, M.S. (1982). Connected algebraic monoids. Trans. Amer. Math. Soc. 272, 693–709.

[70] Putcha, M.S. (1981). The $\mathcal{J}$–class structure of connected algebraic monoids. J. of Algebra 73, 601–12.

[71] Putcha, M.S. (1981). Linear algebraic semigroups. Semigroup Forum 22, 287–309.

[72] Putcha, M.S. (1983). A semigroup approach to linear algebraic groups. J. of Algebra 80, 164–85.

[73] Putcha, M.S. (1984). Reductive groups and regular semigroups. Semigroup Forum 30, 253–61.

[74] Putcha, M.S. (1983). Idempotent cross–sections of $\mathcal{J}$–classes. Semigroup Forum 26, 103–9.

[75] Putcha, M.S. (1983). Determinant functions on algebraic monoids. Communications in Algebra 11, 695–710.

[76] Putcha, M.S. (1986). A semigroup approach to linear algebraic groups II. Roots, J. Pure and App. Alg. 39, 153–63.

[77] Putcha, M.S. (1985). Regular linear algebraic monoids. Trans. Amer. Math. Soc. 290, 615–26.

[78] Putcha, M.S. (1986). Finite semigroups associated with linear algebraic monoids. Quart. J. Math. 2, No. 37, Oxford, 211–19.

[79] Putcha, M.S. (1986). A semigroup approach to linear algebraic groups III. Buildings, Canadian J. Math. 38, 751–68.

[80] Putcha, M.S. (1984). Regular algebraic monoids. in Proc. Marquette conference on semigroups, 183–96.

[81] Putcha, M.S. (1983). Matrix semigroups. Proc. Amer. Math. Soc. 88, 386–90.

[82] Putcha, M.S. (1983). On the automorphism group of a linear algebraic monoid. Proc. Amer. Math. Soc. 88, 224–26.

[83] Putcha, M.S. (1984). Algebraic monoids with a dense group of units. Semigroup Forum 28, 365–70.

[84] Putcha, M.S. (1987). Conjugacy classes in algebraic monoids. Trans. Amer. Math. Soc. 303, 529–40.

[85] Putcha, M.S. (to appear). Monoids on groups with BN–pairs. J. Algebra.

[86] Putcha, M.S. (to appear). The local semilattice of chains of idempotents.

[87] Putcha, M.S. (to appear). Algebraic monoids whose non–units are products of idempotents. Proc. Amer. Math. Soc.

[88] Putcha, M.S. (to appear). The monoid generated by projections in an algebraic group.

[89] Putcha, M.S. and Renner, L. (to appear). The system of idempotents and the lattice of $\mathcal{J}$–classes of a regular algebraic monoid. J. Algebra.

[90] Renner, L. (1978). Automorphism groups of minimal algebras. Masters Thesis, University of British Columbia.

[91] Renner, L. (1982). Algebraic monoids. Ph.D. Thesis, University of British Columbia.

[92] Renner, L. (1983). Cohen–Macaulay algebraic monoids. Proc. Amer. Math. Soc. 89, 574–78.

[93] Renner, L. (1984). Quasi–affine algebraic monoids. Semigroup Forum 30, 167–76.

[94] Renner, L. (1985). Reductive monoids are von Neumann regular. J. of Algebra 93, 237–45.

[95] Renner, L. (1985). Classification of semisimple rank one monoids. Trans. Amer. Math. Soc. 287, 457–73.

[96] Renner, L. (1985). Classification of semisimple algebraic monoids. Trans. Amer. Math. Soc. 292, 193–223.

[97] Renner, L. (1986). Analogue of the Bruhat decomposition for algebraic monoids. J. of Algebra 101, 303–38.

[98] Renner, L. (to appear). Conjugacy classes of semisimple elements and irreducible representations of algebraic monoids. Comm. in Algebra.

[99] Renner, L. (to appear). Representations of rank one algebraic monoids. Glasgow J. Math.

[100] Renner, L. (to appear). Class groups of semisimple rank one monoids. Proc. Amer. Math. Soc.

[101] Renner, L. (to appear). Connected algebraic monoids. Semigroup Forum.

[102] Renner, L. (to appear). Completely regular algebraic monoids. J. Pure and Applied Alg.

[103] Rhodes, J. (1986). Infinite iteration of matrix semigroups II. Structure theorem for arbitrary semigroups up to aperiodic morphism. J. Algebra 100, 25–137.

[104] Schein, B.M. (1968). Strongly regular rings, in Summaries of talks of the 1st All–Union Symposium on the theory of rings and modules. Kishinev, 38–9.

[105] Schein, B.M. (1972). Pseudo–semilattices and pseudo–lattices. Izv. Vyss. Ucebn. Zaved. Mat. 2(117), 81–94. English Transl. in Amer. Math. Soc. Transl. (2)119.

[106] Shafarevich, I.R. (1977). Basic algebraic geometry. Springer–Verlag, New York.

[107] Sizer, W.S. (1980). Representations of semigroups of idempotents. Czechoslovak Math. J. 30(105), 369–75.

[108] Springer, T.A. (1983). Linear algebraic groups. Birkhäuser, Boston.

[109] Srinivasan, B.R. (1968). Weakly inverse semigroups. Math. Ann. 324–33.

[110] Steinberg, R. (1974). Conjugacy classes in algebraic groups. Lecture Notes in Math. 366, Springer–Verlag, New York, 1974.

[111] Suzuki, M. (1982). Group theory I. Springer–Verlag, New York.

[112] Tamura, T. (1972). On Putcha's theorem concerning semilattice of archimedean semigroups. Semigroup Forum 4, 83–6.

[113] Tamura, T. (1975). Quasi–orders, generalized archimedeaness, semilattice decompositions. Math. Nachr. 68, 201–20.

[114] Tamura, T. and Kimura, N. (1954). On decomposition of a commutative semigroup. Kodai Math. Sem. Rep. 4, 109–12.

[115] Tits, J. (1974). Building of spherical type and finite BN–pairs. Lecture Notes in Math. 386, Springer–Verlag, New York.

[116] Waterhouse, W.C. (1982). The unit groups of affine algebraic monoids. Proc. Amer. Math. Soc. 85, 506–8.

[117] Winters, D.J. (1982). The fitting and Jordan structure of affine semigroups, in Lie algebras and related topics. Lecture Notes in Math. 933, Springer–Verlag, New York.

[118] Yoshida, R. (1963). On semigroups. Bull. Amer. Math. Soc. 69, 369–71.

[119] Zalcstein, Y. (1973). Locally testable semigroups. Semigroup Forum 5, 216–27.

INDEX